CRUSADER
JOHN COBB'S ILL-FATED QUEST FOR SPEED ON WATER

For John Rhodes Cobb
"If it goes wrong at these speeds? I've just stopped thinking about it!"

And for Mark Harley
Goodbye old friend.

CRUSADER

JOHN COBB'S ILL-FATED QUEST FOR SPEED ON WATER

STEVE HOLTER

© Steve Holter 2021

All rights reserved. No part of this publication may be reproduced or stored in a retrieval system or transmitted, in any form or by any means, electronic, mechanical, photocopying, recording or otherwise, without prior permission in writing from Evro Publishing.

Published in May 2021

ISBN 978-1-910505-61-8

Published by Evro Publishing
Westrow House, Holwell, Sherborne, Dorset DT9 5LF, UK

www.evropublishing.com

Edited by Mark Hughes
Designed by Richard Parsons

Printed and bound in Malta by Gutenberg Press

Front cover
Commissioned by the author, this work by Dutch artist Arthur Benjamins portrays *Crusader* at speed on Loch Ness in September 1952.

Rear cover
Computer renditions by Mick Hill of the general arrangements of *Crusader*, produced from the original drawings and in-depth research of the few colour photographs available.

Frontispiece
Photographed from Vosper's 'jolly boat', John Cobb sits on *Crusader*'s cockpit surround as she is towed in after a run. *John Bennetts, courtesy of Julie Newton*

CONTENTS

Foreword · Ken Warby — 6
Foreword · Sally Railton Joslin — 7
Preface — 11
Acknowledgements — 15
Introduction — 17

1 Quiet Giants — 19
2 From Wheels to Water — 45
3 Fastest Man on Earth — 63
4 Reversing the 'Three Pointer' — 73
5 Tensions Rise — 89
6 Slow Going — 137
7 A Line in the Sand — 165
8 Construction Begins — 182
9 On Scottish Waters — 200
10 The End — 240
11 Why? — 253
12 From the Ashes — 288
13 A Point to Prove — 316

Postscript — 328
Appendix — 339
Index — 347

FOREWORD
KEN WARBY
MBE

Few men have gone very fast on water, and got away with it! I built *Spirit of Australia*, in my backyard, from wood and using hand tools, and a jet engine. I didn't look back to Cobb and *Crusader*, I looked forward at what I knew through my own powerboat racing and building experience.

Many really fast boats have been constructed with wood. It's actually one of the best things to use, cheap, easily modified, and flexible, and years of experience with designing, building and racing wooden race boats, wood has been proven over many decades in powerboat racing from limited-class racing through to unlimited hydroplanes. I've designed, built and driven a wooden boat to speeds over 300mph, I can attest to its use.

When Steve let me read the original manuscript for *Crusader*, my reaction was one word: "Superb!" Sure, they were working in the dark, it was new, but without seeing all that went behind it, it would have been easy just to think they got it wrong. They didn't, but mistakes were made, and we know that because nearly 14 years of research and hard work by Steve has made history speak, it's like sitting in the pub earwigging on a conversation and unable to join in and debate the mistakes.

The water speed record is a very dangerous sector, but I have always said, you can't go wrong with a well-designed and well-built boat. *Crusader* was only half of that. Steve has given us a chance to see this for ourselves.

And now, I'm at it again with my son, in a boat made of wood, with a jet engine, chasing the unlimited world water speed record.

FOREWORD
SALLY RAILTON JOSLIN

The memory of sitting with my father in his darkened room watching film footage of the last run of *Crusader* after he returned from Scotland in 1952 remains clear to this day. The screen and projector had been carefully arranged, my father's battered armchair (covered in burns from his pipe embers) had been placed nearby, and the foot stool he shared with the respected family cat "Boots" had been politely vacated for me. With graph paper, pencils, and his well-worn "stop watch" in hand, the investigation of the various films sent from the UK commenced. Little was said during the running of these films, other than that it was our task to determine exactly when *Crusader* started "bobbing" during the early stages of that tragic run, and to freeze-frame, enlarge, and measure the up/down motions so they could be plotted on graph paper.

From this period onwards the subject of *Crusader* was only mentioned once more, when my father and I drove up to Inverness to "have a look around". And in answer to my question of what he thought had caused the accident, he said he had a fair idea, but as the evidence was now at the bottom of Loch Ness and unlikely to be found, this question would remain unanswered.

When my father died in 1977, my brother Tim went through his boxes of papers, correspondence, photos, and "unrecognizable doodles", and sent me what he deemed worthwhile, as he didn't wish to keep any of it himself. On receipt I had a marathon read through, and was completely captivated; but with a growing family and a busy life I packed it all up and "filed" it under a bed, where it languished for the next 30 years.

In 2005 a plan to make a documentary about *Crusader* surfaced, and I was put in touch with John Cobb's delightful niece Diana Sweeney. She had met the group involved in the project and was concerned about their views

CRUSADER

Reid Railton was photographed by his daughter Sally during a trip to Britain in 1957, when Riley loaned him one of its new One-Point-Five models in order to find out his impressions of it. *Sally Railton Joslin*

of the story, and urged further investigation if I cared to invest the time in it. The first thing I did was to pull out the correspondence between John Cobb, my father, and Peter Du Cane (who was the builder of the boat), which fairly quickly confirmed her concerns. And over the next several years she and I had many useful and humorous chats, when she was also a pillar of support during some challenging times on the way to discovery.

During this time I was fortuitously directed to Steve Holter, who was reported to independently know a great deal about *Crusader*'s history. This was the turning point in getting this story accurately recorded for posterity. I will be forever grateful to Steve for so diligently and passionately seeking out all the pieces that make up the whole, as now reflected in this book. It is hard to describe today the relief I felt when first speaking with him about these

FOREWORD · SALLY RAILTON JOSLIN

matters, and the strong respect he held for the truth, whatever that turned out to be. Steve never allowed himself to take anything for granted, and tactfully curbed my thinking when I overreached.

As Steve's project was coming into focus, I was encouraged to consider a more comprehensive biography of my father, and was honored to learn that Karl Ludvigsen would be interested in tackling this rather large-scale undertaking. In due course this became the marvelous two-volume presentation called *Reid Railton: Man of Speed*, which won six prestigious book awards and was also published by Evro. For this biography, Steve provided much of his research into *Crusader*, and Karl most skillfully included many aspects of it within his narrative.

Steve has not only gathered an enormous amount of documentation over recent decades, but his continued sleuthing has helped lead to some new discoveries. Most significantly, *Crusader* has been discovered on the bed of Loch Ness and photographed by *National Geographic* using state-of-the-art technology. As well, Steve located a scale model of *Crusader*'s potential successor and this is now under the enthusiastic stewardship of Richard Noble, who is developing a working version in order to find out more. One can only wonder how all this may evolve.

I am sure my father would be highly amused by all these "goings-on" — and also exceedingly interested to know about the various revelations that have emerged from Steve's research.

PREFACE

I have spent a considerable part of my life reading, researching and writing about engineering, cars, motor racing and record breaking, as a racer, museum curator, engineer, television researcher, archivist and lecturer. I find it too easy to become engrossed in digging through old files, photos and paperwork, but sometimes a change is as good as a rest. Some 17 years ago, I was at that point where I wanted a project and wanted to write something, but I liked the idea of a change.

I had been approached with several suggestions to write a biography — a well-known 1960s disc jockey, a comedian and impersonator, an iconic musician who had held quite a fascination for me — but the lack of new detail made me think I could not come up with much that had not been written before.

Then I received an email from my old friend, master modeller Fred Harris, who told me that someone called Sally Joslin was trying to contact me. When we finally spoke, she introduced herself as Sally Railton. The penny dropped.

In all that I had read about Malcolm and Donald Campbell, Goldie Gardner, ERA racing cars, Brooklands and car design in general, the name Reid Railton had come up more than a few times. He was the design genius of his time.

As Sally and I chatted, we began to see the possibility of a full biography, but later we agreed that the subject required a specialist writer. Eventually that project went to Karl Ludvigsen and a fine job he did in his book *Reid Railton: Man of Speed*. For my part, I preferred a bit more oil and grease, and took on the thorny issue of Railton's last big project, *Crusader*, John Cobb's ill-fated, jet-powered hydroplane. As I researched this part of the jigsaw puzzle for Karl, it became very apparent from conversations with Sally, members of the Cobb family and others who were involved and were still alive, as well as from my old notes and interviews with some who were no longer with us, that here was a story on its own. I found myself once again writing about engineering, men and machines, and the tragic final chapter for one of Britain's most unsung kings of speed.

Invariably, when researching a subject, there comes a time when one believes that all available information has been found, and there is nothing new to read, hear or see. However, just after completing the first draft of this book, I tracked down a film that had not been viewed for 47 years. There had never been any previous written or verbal reference to it, and, at the time of writing, its origin remains a mystery.

Two aspects of this film make it unique. One is the access to the *Crusader* project that must have been afforded to the cameramen (there was clearly more than one), who were not professionals judging by the framing of some shots, although the contents show that it is no mere 'home movie'. The other unique feature is the key elements that are captured, such as the modifications to the forward planing shoe and its rear braces, the crowds that blocked the A82 on the day of *Crusader*'s last run and, sadly, the recovery of Cobb's body after his accident.

This film also confirms that misreporting of events at the time, and subsequent repetition of this flawed reporting, leads to history being incorrectly recorded, so that the truth becomes hidden in the mists of time. This film, along with recently discovered archive material, has given me an unexpected opportunity to correct this, and this book takes full advantage of having history 'as it happened' to draw upon. As an example, it has always been written that spectators assumed that Cobb had been merely injured in the crash that befell him, that he sat upright in the recovery boat, and that he died later. Regrettably, that simply was not the case, as this 'new' footage harrowingly portrays. I am grateful to the Moving Image Archive and the staff of the National Library of Scotland for making it possible to see these events 'first hand'.

When it came to the 'paper trail', Sally Joslin, the families of John Cobb, Peter Du Cane and George Eyston, as well as BAE Systems (Vosper Thornycroft), the Hampshire Archive and Castrol, made it possible to build a triangle of communication between the project's key players. This was probably the first time all three components had ever been put together as a single entity, and I was certainly the first person to read all of it. Even I have to admit that some aspects of my previous writings about *Crusader* had not been entirely accurate, because prior to my research no one had ever had access to all three major archives, or the new material that I found languishing in some dark and dusty corners. I merely had to put everything together in chronological order, and let Reid Railton, John Cobb and Peter Du Cane tell the story — and then readers can form their own opinions.

PREFACE

There will be some who will not like parts of this account — but there is no conjecture and this is history in the words of those making it.

And so here is the true account of the *Crusader*, the world's first purpose-designed and purpose-built jet-propelled boat. *Crusader* was based on concepts that were years ahead of their time and she was built in an early post-war period in which materials were scarce and staid conservatism still characterised the world of boat construction. The entire project was barely able to keep up with one man's advanced thinking.

That it all ended so badly is a matter of history. But as to why it ended so badly — that is for you to decide.

If I have learned anything from writing this book, it is that when we master our problems, and life runs the risk of becoming mundane, the only antidote is the pursuit of new goals.

ACKNOWLEDGEMENTS

It is impossible to write a book, any book, without help. I am indebted to many generous, knowledgeable, informed and talented people. Some I am lucky to have had as friends for many years, while others, although not acquaintances, felt they could share whatever they could. Some I was fortunate to know before they finally took their memories from this world, after sharing them, but they all made a thoroughly enjoyable task that much easier. All of them deserve public thanks, for providing anything from photos and recollections, to encouragement and friendship, or just inspiration. Some are named here, in no particular order, but there are always many others who contributed in some way and they will know who they are. I thank you all.

Without the 'kick start' and encouragement I received from Sally and Jim Joslin, who shared so much information and time with me, it is unlikely that I would have put finger to keyboard, but all the names that follow have also ensured, in one way or another, that the story is now there for all to see. Two old friends, Fred Harris and Kevin Desmond, also helped to set the ball rolling. Then comes a very long list: George, Olga and Basil Eyston; Bert, Richard, Rodney and Paul Denly; Karl Ludvigsen; Charles Taylor; Steven Parry-Thomas; Vera 'Vicki' Cobb; Nick Mouat; the family of Margaret Cobb; Charles Du Cane; Adrian Shine and the Loch Ness Project; Leo Villa; John Dyson; Ady Simpson; Ken and Lewis Norris; Rich Marsh; Lex Lambert; Howard Statham; Ollie Furey; Diana 'Bunny' Sweeney and her son Theo Mezger; Doug Nye; David Tremayne; Gary Critcher; Malcolm Pittwood; Colin and Ricky Cobb; Gino Felicetti; Cyril Posthumus; Rebecca Leppard; David Weil; Geoff Holden; Michael C. Feltham; Sir Stirling Moss; Stuart Dent; Colin Torode; L.G. Brown & BAE Systems (formerly Vosper Thornycroft); Brooklands Museum; Westland Aerospace (formerly Saunders Roe); Mike Evans and Rolls-Royce Heritage Trust; Zillwood 'Sinbad' Milledge; Richard Setchim of Price Waterhouse; records holders of Elders Anning Ltd (formerly Anning, Chadwick & Kiver Ltd); Ken and Dave Warby; Geoff Holme and Frank Lydall; The Hampshire Archive; Reg Beauchamp; the family of Ted Cope; Ray Baxter; Barry Stobart-Hook; Sally Durlac (Cobb family descendant); John Hawtyn Cobb; the family of George 'Guy' Bristow;

CRUSADER

the estate of Hugh Jones; Bill Maloney MRINA; Ken Wheeler EUR ING CEng MIMechE FCMI CDipAF (ACCA); Ray Kleboe; the estate of Harry Cole; Nick and Sarah Stevens; Mrs U.E. Goldie Gardner; The Procurator Fiscal for Inverness and the Scottish Fatalities Investigation — North; Gordon Menzies; Roberta Simpson and all at the National Library of Scotland (Moving Images Archive); Lorne Campbell; Richard Noble; Joanne Burman and Andy Griffin of Castrol Burmah and the BP Archive; Craig Wallace and Kongsberg Maritime; John 'Lofty' Bennetts and his son-in-law and daughter Len and Julie Newton; Mark Hughes and Evro Publishing; David Haffner; master modellers Ernie Lazenby and John May.

Special thanks to my old friends Arthur Benjamins (for the front-cover artwork), whose website www.pilgrimstudio.com I urge you to visit for prints of this and other works of automotive art, David de Lara for crash analysis composites and stills, and Mick Hill for the superb computer renditions.

And finally, Charlie and Ross Holter, and Deb, who despite having no interest in the subject, got to hear a lot about it!

INTRODUCTION

"This is to certify that Mr John Cobb, piloting *Crusader* registered #6, established the following figures on the statute mile on Loch Ness on September 29th, 1952. Elapsed time: 17.4 seconds. Speed: 206.89 statute miles per hour. This performance of one run over the mile does not constitute a World's Record, but it is the fastest speed ever obtained on water."

The words above are from the official statement made by the UIM (Union Internationale Motonautique) to mark the tragic end of John Cobb's attempt to beat the water speed record held by American Stanley Sayres and return it to Britain. The UIM also made the unique gesture of issuing an official certificate even though the tragedy occurred during the first of *Crusader*'s two required runs and therefore Cobb's speed on that run could not be a new record.

Those 61 words were the culmination of four years of design and research in a joint venture between Cobb, his long-time friend and designer Reid Railton, and boatbuilder Vosper Ltd under the management of Commander Peter du Cane.

Cobb already held the land speed record, taken in a car designed by Railton, but in deciding to follow Henry Segrave and Malcolm Campbell from land to water, both men realised that they needed the skills of an experienced boatbuilder — and so an uneasy coalition came into being.

Vosper had already constructed Sir Malcolm Campbell's successful *Blue Bird K4*, but that was a craft built in traditional materials, in a traditional way. Railton's fertile imagination conceived something quite out of the ordinary, a revolutionary design that not only used jet propulsion but threw standard hydroplane practice out of the window by reversing the three-point layout, thereby introducing the canard hydroplane.

It was more rocketship than boat. It was both prototype and finished article all in one. The entire concept was new territory and all concerned were travelling into the unknown.

To Railton it was food and drink, and Cobb was more than happy to "go along with Reid's genius". To Du Cane and Vosper, however, it was a step too far from tradition, and the project ended in disaster on a distant loch in Scotland four years after it began.

The story of *Crusader* appears to be a straight-forward tale of two men's ambition to break the water speed record only for tragedy to strike at the last hurdle. But when the events and characters involved are studied in depth, and note is made of the period in which this all took place, it soon becomes obvious that the story is more complex and interwoven. The journey that ended on 29th September 1952 is best told in its component form, then assembled like one of Reid Railton's creations.

CHAPTER 1
QUIET GIANTS

The public face of the *Crusader* project was that of John Cobb, racing driver, record breaker and businessman, a large and imposing figure physically but also a shy and humble person. His main support was his long-time friend, designer and confidant, Reid Railton, a man of slight build and a deep thinker for whom problems existed solely to be solved.

Both of these men of speed did what they did because they relished the challenge and excitement of succeeding in the most difficult circumstances. In upbringing, background and outlook, they were the perfect foils for each other, quiet giants who were happy to pursue their quest for speed calmly and unobtrusively — and let others bask in the limelight.

Reid Railton

There is an adage about not needing a sledgehammer to crack a nut. This could apply to Reid Antony Railton. Where some would look at a problem and seek a solution head-on with a bludgeon, Railton had an innate sense of engineering and would adopt a more precise approach, with a scalpel. His methodology never wavered, regardless of whether a problem was large or small, or if the available motive power was massive or miniscule, or if the obstacles were major or minor. What others would achieve with force, Railton would arrive at with subtlety. Railton would never use complex hydraulics if a job could be done with a spring and a lever. He would never let go of a problem once he had begun to tackle it.

To some, he was 'merely' an automotive engineer, a designer of vehicles to break speed records, a deep and obsessive thinker. But that is to totally understate the influence his skill and knowledge has had on engineering and design, an underestimation exacerbated by his overwhelming desire to keep work separate from family, to either blend in with a team or remain at home in private.

His philosophy was simple: something was either right or wrong. He never minced words about anything, least of all about a driver. He could be seemingly sarcastic, although every word he uttered was always absolutely accurate. Usually, though, Railton was not a 'talker' and social ambitions were never a priority. In particular, it was difficult to get him to express an opinion about the designs of others, especially when in the company of engineers. He did not have a robust constitution and often, perhaps through deep concentration, developed severe headaches that he would seek to cure by banging his forehead harshly with a hand.

Railton was very much 'a backroom boy', with much of the apparent vagueness a brainy scientist is supposed to exhibit — yet he was very far from being vague about engineering and design. He possessed an impish sense of humour that could tempt the unwary into arguments, from which they could emerge feeling rather foolish and without quite understanding why. Given a problem, even under difficult circumstances when creating a machine for racing or record breaking, he would revel in complex calculations from which something quite stunning in its apparent simplicity would result. Because of his thoroughly logical approach to problem solving, he was not one for readily believing unsupported 'facts' aimed solely to impress. He would curtly dismantle 'false truths' with resolute precision, usually while trying to save face for those who had proposed them.

For all this, he managed to keep everything outside of design and engineering to himself and those close to him, not once seeking to promote himself in any way.

Reid Railton as a schoolboy. Noted by his teachers as a studious pupil, he was already fascinated by the workings of all things mechanical. *Railton archive*

QUIET GIANTS

Railton was born on 24th June 1895, in Chorley, Alderley Edge, Cheshire, and christened on 13th August. His father, Charles Withington Railton, was a Manchester-based stockbroker and his mother, Charlotte Elizabeth (*née* Sharman), was — in the parlance of the day — a 'home maker' who spent much of her spare time supporting many local causes and charities. He was educated at Rugby School and then Manchester University. After graduation in 1917, he joined Leyland Motors in Lancashire to work under J.G. Parry Thomas, who proved to be the perfect tutor, in both engineering and life, and was to become the glue that would fix together the lives of Reid Railton and John Cobb.

John Godfrey Parry Thomas — who was called 'John' by those who knew him but 'Parry' by those who did not — was the son of a vicar, born in Wrexham on 6th April 1884. At the age of five he started at Oswestry School, where he is said to have raffled his sixpence-a-week pocket money at a penny a go, often making a useful profit. He became fascinated with engineering from an early age and in 1902 went to study electrical engineering at the City & Guilds Engineering College in London. There he met and became friends with Ken Thomson, who would become his assistant at Brooklands and would later set up Thomson & Taylor, the famous Brooklands-based specialist engineering company that would develop many successful racing and record-breaking cars.

By 1908, Parry Thomas had already designed an infinite-ratio electrical transmission system, which proved too advanced for the materials and manufacturing techniques of the time but within 70 years was commonly fitted to passenger-carrying vehicles. He also came up with the principle of a self-locking threaded nut but was again thwarted by the materials of the period, although the concept later became the 'Nyloc' nut. It was this sort of advanced thinking that led to demand for his services from government advisory boards during the First World War.

After numerous jobs, Parry Thomas was taken on as chief engineer at Leyland Motors, just as the company was investigating the concept of creating the perfect car, one that would compete with the benchmark of the time, Rolls-Royce, in both luxury and performance. Parry Thomas was the new car's designer, helped by his young protégé Reid Railton. Dubbed the 'Lion of Olympia' after it was first shown at the 1920 motor show in London's Olympia, the magnificent Leyland Eight was an extremely expensive beast with a basic price of £2,700 (equivalent to over £120,000 now). Although Leyland's investment was huge, orders were slow and eventually only 14 cars were built, including two for the Maharajah of Patiala and one for Michael Collins, the Irish revolutionary, soldier and politician.

Each Leyland Eight was tested to 100mph by Parry Thomas after fettling and tweaking, with Railton's help, to ensure that this remarkable velocity — the land speed record at that time was only 124mph — could be achieved. Parry Thomas was crestfallen by the Eight's commercial failure but so convinced of the car's potential that in 1922 he decided to go motor racing with one. The logic was simple: good race results with a standard road car would prove the excellence of the design. His plan, however, was a step too far for Leyland Motors and the company refused to support the venture. Parry Thomas's belief was such that he decided to go off on his own, leaving Leyland and heading south to make his way in racing as an engineer and driver.

Brooklands, near Weybridge in Surrey, was the first purpose-built motor racing circuit in the world, created as a proving ground and showcase for the British motor industry. It was almost inevitable that Parry Thomas would gravitate there, eventually even living and working within the confines of the track itself in a bungalow, 'The Hermitage', which had been built during the First World War, probably as a residence for Royal Flying Corps officers.

Documentation within the Leyland archive suggests that Railton played an important role in ensuring that the split between Leyland and Parry Thomas was amicable, and, under the guise of conducting 'secret' development work for Leyland, he arranged for his mentor to receive several chassis and a quantity of spares from the factory. Soon afterwards, Railton left Leyland as well, setting up the Arab Motor Vehicle Company and taking the role of chief designer.

Always hands on, John Parry Thomas makes adjustments to the 7-litre straight-eight engine of his Leyland Eight at Brooklands in 1923. Note that the car, although converted for racing, retains its horn, which its driver was not above using on track. *Getty Images/ Topical Press Agency*

Parry Thomas, with Railton, about to set off around Brooklands. Not only were the pair involved in the design, manufacture and fitting of all modifications, but they tested them as well. *Author's collection*

The Arab car came about following talks between Parry Thomas, Railton and Henry Spurrier, chairman of Leyland Motors. Leyland had produced 50 four-cylinder engine blocks intended for a fast delivery van, but the project had been shelved. The three men discussed what to do with these surplus 2-litre engines and agreed to build a new sports car to use them up. A prototype was constructed based on an Enfield-Allday chassis and taken to Brooklands for the 1924 Easter Meeting. The engine featured an overhead camshaft and the unusual leaf-style valve springs found on Parry Thomas's Leyland Eight design. Drive to the rear wheels came via a Moss four-speed gearbox with an ENV spiral-bevel rear axle. The Arab was one of the first English cars to have an electric fuel pump. Production started in 1926 at a factory in Letchworth, Hertfordshire, and two versions were offered — a low-chassis sports model and a standard-chassis touring car. Two- and four-seater bodies were available on the tourer, which cost £525, but only the two-seater on the sports chassis, priced at £550.

Parry Thomas set up a design room and workshop in 'The Hermitage', the home that he shared initially just with his Alsatian dogs, Togo and Bess, plus a housekeeper and her husband along with their children when they came home during school holidays. A little later, his friend Ken Thomson also come to work there as Thomas's assistant and together they formed the Thomas Inventions Development Co. Ltd. In due course Railton arrived as well, at first merely as a regular visitor from the Arab works, acting as a consultant to Parry Thomas on an *ad hoc* basis, but eventually joining full-time.

The Thomas/Railton/Thomson trio, later joined by Reginald Beauchamp, developed the Leyland Eight into a fearsome racing machine, the *Leyland-Thomas*. With its engine enlarged to over 7 litres, the car achieved success at Brooklands on the famous Outer Circuit. Railton then began to think that perhaps a more subtle approach might produce an equally successful Outer Circuit car and work started on the first of the 'Thomas Flat Iron Specials'. For Parry Thomas's part, he had developed a liking for big, large-capacity cars, and had been taking an interest in one called the *Higham Special*, built and developed by Count Louis Zborowski, a prominent racer of the time. Named after Zborowski's estate in Kent, the *Higham Special* was a massive machine powered by a 27-litre Liberty aero engine. After Zborowski's death during the Italian Grand Prix at Monza in October 1924 when his Mercedes hit a tree, Parry Thomas bought the *Higham Special* for a mere £125.

Parry Thomas's intention was to develop the *Higham Special* into a potential land speed record contender, thereby proving the skills of his group of engineering renegades. He decided upon the land speed record not because of the element of speed but because of his passion as a dedicated engineer for research and progress, a trait that Railton mirrored. Parry Thomas was certainly not after publicity as he was another shy, retiring man who avoided the limelight. He also rarely smiled as he was self-conscious about his prominent front teeth.

On the face of it, Parry Thomas's acquisition of the *Higham Special* was rather strange. It used a one-off Rubery Owen chassis and its Liberty engine was mated to a pre-war Benz gearbox that drove the rear wheels via archaic twin chain drives. The entire concept was not exactly 'state of the art' and could be described as 'just short of Heath Robinson' — but it suited Thomas's purpose. As for Railton, however, the *Higham Special* was anathema to him: he considered it to be brute force mated to agricultural machinery.

Parry Thomas christened the car *Babs* and set about an extensive programme of work. Considerable weight was lost by removing unwanted chassis components and replacing some cast-iron fittings, and Railton began the process of looking for more subtle gains and improvements that could be made. In October 1925, Parry Thomas took *Babs* to Pendine Sands, South

Parry Thomas in his trade-mark Fair Isle pullover steers Babs as she is pushed out onto the sands of Pendine for a run. Later he would run the drive chains much tighter than in this photo. *Alamy/Mirrorpix*

Wales, for his first attempt on the land speed record, which Malcolm Campbell had established at 150.87mph in his Sunbeam 350HP three months earlier at the same venue.

Parry Thomas and *Babs* arrived at Pendine Sands just as winter was setting in and unfortunately poor weather precluded any chance of a record-breaking run. Back at 'The Hermitage', Railton, Thomson and Beauchamp presented him with a five-page job sheet of further modifications that they thought would improve *Babs*. Beauchamp recalled, "John [Parry Thomas] had driven himself back to Brooklands from Wales, he looked worn-down and tired, but he scoured the rather tatty sheets, shrugged, said 'Right boys, are you fit?' and began to roll up his sleeves."[1]

By April 1926, the improved *Babs* was back at Pendine, this time with funding from Shell-Mex, for another attempt at the record, which Henry Segrave now held after attaining 152.33mph at Southport Sands, Lancashire, the previous month. After just a couple of warm-up runs, Parry Thomas set a new record of 169.30mph, and the next day, 28th April, he increased it to 171.02mph.

Following that achievement, the rest of 1926 was just as successful for the

CRUSADER

Parry Thomas and *Babs* at Pendine in 1926, the year the pair took their last record at 171.02mph. John appears relaxed enough at speed to take a look at the camera. *Getty Images/E. Bacon*

The sad remains of *Babs* sit on the sand after Parry Thomas's fatal crash on 3rd March 1927. His friend and protégé Reid Railton determined that the right-hand wheel collapsed and an errant spoke went into the adjacent sprocket, causing the very tight chain to snap. *Private collection*

Brooklands-based quartet, with Parry Thomas winning numerous races at Brooklands and in October adding a further eight 'class' speed records there in other cars from his racing stables. Yet, in among the 12 various records established that year, ranging from outright land speed to class and endurance, both Parry Thomas and Railton found time to offer advice to Malcolm Campbell and his land speed team as they built the first *Blue Bird*. So began the consultancy work that was to become the cornerstone of the Thomas/Railton/Thomson/Beauchamp group, activity that Railton would continue for the rest of his life. Indeed, it was Railton who was taken on to sort out the troublesome and very complicated Joseph Maina-designed gearbox in *Blue Bird*. Campbell's first foray with his new car came early in 1927 at Pendine Sands, where he recorded a new record of 174.883mph on 4th February.

That same month, Henry Segrave was on his way to Daytona Beach, Florida, USA, to go for the record in his new twin-engined '1000HP Sunbeam', which had been built so secretly that it was often referred to as the 'Mystery Racer'. As the Sunbeam was expected to break the 200mph barrier with some ease, Parry Thomas took *Babs* — in the firm knowledge that 200mph was far beyond his car's reach — back to Pendine in an effort to raise the record before Segrave could even run his new car.

Parry Thomas was not a well man at this time. He had been working intensely, often at night as well as all day. He had 'flu, and the cold, damp conditions at 'The Hermitage' had not been good for his health. He was a big man, in every sense, and normally weighed nearly 17 stone, but by the time he arrived at the Beach Hotel at Pendine he looked thin and his face was grey and drawn. Railton beseeched him to wait, but Thomas was also a determined man. For this attempt, he had additional assistance from Dunlop, which had started preparations soon after Thomas's previous Pendine record.

On 3rd March, *Babs* was ready and conditions were favourable. After the usual start and warm-up procedures, Parry Thomas set off on a timed run. Midway through the seven-mile course, the car lurched and then slewed around in a large arc before coming to rest with flames taking hold. When Thomas's crew reached the scene, they found him dead in the car, partially decapitated and with deep cuts to his neck. In order to recover his body before the fire worsened, they had the unpleasant task of breaking his legs to pull him clear.

The unenviable job of working out what had gone wrong fell to Reid Railton, who had remained at Brooklands and learned of the tragedy by telephone. The moment he put down the telephone on Thomas's desk in 'The Hermitage', he set off for Pendine to examine the remains of *Babs*. He noticed a dent in the steel sprocket on the right-rear wheel and observed that a spoke from the wheel fitted the dent perfectly. From this, he deduced that the wheel had collapsed, and a spoke had become trapped between the chain and the drive sprocket. This was enough either to stretch and snap the chain or at least dislodge it. From the angle and position of Thomas's injuries, and the damage to the right side of the car, it could be determined that the chain had flailed around as *Babs* righted herself, hitting Thomas's head.

It was Railton who instructed that *Babs* should be buried there on the beach. Some suggested it was an emotional decision, but it is more likely that Railton simply chose the pragmatic option, realising that the cost of recovery and repair would be too high, and that the wreckage would serve as a fitting memorial if left where it was. Ironically, Railton and Beauchamp had just completed drawings for a new gearbox and drive-shaft conversion to replace the chain drive.

With Parry Thomas gone, Ken Thomson was now joined by Ken Taylor and they continued the activities of the Thomas Inventions Development Co. Ltd. under the new name of Thomson & Taylor. Reid Railton stayed on as technical director and chief designer, typically refusing to have his name appear on any letterhead or signwriting.

Meanwhile, Railton lost interest in the Arab concern and closed the Letchworth works after only six or seven cars had been made. Thomson & Taylor bought the remaining components and assembled another five cars. The Arabs had impressive performance, the sports car being capable of an easy 80mph while the Super Sports could exceed 90mph. At least one touring chassis was supercharged and is said to have achieved over 105mph.

Railton also had considerable input into the design of the Riley Nine Brooklands, the prototype of which had been prepared to compete in the BARC (Brooklands Automobile Racing Club) meeting in August 1927, with Parry Thomas having been the intended driver. Instead, Railton stepped in to race the car in the '90mph Short Handicap' and won by a country mile, lapping at a remarkable 98.62mph. Railton had never raced before and would not do so again.

Thomson & Taylor prospered. Ken Taylor was a chip off a very different block from Ken Thomson and Reid Railton. He was a craftsman who had risen to a management role and knew how to run a business. It fell to him to see out Railton's ideas and designs, and subsequently nurse the resultant

Reid Railton (left) stands with Malcolm Campbell at Thomson & Taylor, Brooklands, 1932. This was the first year in which *Blue Bird* used the Rolls-Royce R-type engine, and here engine number R33 can be seen connected to the air starter for its first run-up in the chassis. *Courtesy of Charles Taylor*

machines through whatever competition they were entered in, and then take care of modifications and repairs needed after the abuse meted out by some of the drivers. Thomson & Taylor's reputation grew ever stronger — and Taylor made certain the company was properly paid for its work.

It was a short walk from Reid Railton's office at Thomson & Taylor to the Vickers Aviation wind tunnel, which Railton was able to use to keep Malcolm Campbell's *Blue Bird* successful in its various land speed record attempts at Daytona Beach, while constantly having to redesign and improve the car's technical specification without spending any more of Campbell's money than was absolutely necessary. Indeed, cost was as important to the canny Campbell as the speeds he wanted to attain. In a letter to Railton regarding the 1935 incarnation of *Blue Bird* and its planned runs at a new venue, Bonneville Salt Flats, he wrote:

> I shall require your assistance with the press release for the forthcoming attempt at 300 miles per hour at Bonneville in September, as you have been there and know what to let the press boys know about running on

salt. Please refrain from numbering the old girl, she is the 'Blue Bird', not the first, nor the second, it's so annoying to see 'Blue Bird VI or V', it shows the lack of appreciation from them of the skilled development work we have done to chassis, engine and body type over the years, a point I shall be making quite strongly. These press things are ideal to improve the revenue, but have to be right.

It was gratifying to hear from you that [Leo] Villa is on the right lines with the supercharger problems on the Blue Bird and I look forward to seeing the results of the wind tunnel tests when you have them, and trust that the charges for this will be minimal as the idea was 'in house', so to speak.[2]

Railton had by now established a reputation that saw him sought out by many, in racing and road spheres, and as such he had fingers in many pies.

He became involved in a design for a luxury road car, as he had before with Parry Thomas and the Leyland Eight. This time he worked with Noel Macklin, who, having sold his interest in Invicta Cars, became the motivating force in the creation of the Railton car. Macklin saw the possibility of using an American chassis mated to a more European style of bodywork and wanted Railton on board, in the full knowledge that his name would add prestige to the image of a fast, sporting car. This in turn led to Railton working with the Hudson Motor Car Company (supplier of the 'Railton' chassis) in Detroit, which was partly why he and his family moved to the United States in 1939.

Railton was also heavily involved in the design and development of the ERA racing car through Thomson & Taylor and its connections with Freddie Dixon, Raymond Mays and Peter Berthon. With backing from Humphrey Cook, Mays and Berthon joined forces to create English Racing Automobiles in a workshop behind Mays's house in Bourne, Lincolnshire, developing their new ERA from the 'White Riley', Mays's supercharged six-cylinder hillclimb car. Although Berthon was an excellent engine designer, the trio knew from experience that if they wanted a light, good-handling chassis, they needed a design from Railton.

The ERAs were built for so-called *voiturette* racing but Mays had ambitions to step up to the Grand Prix arena, where the cars were built to different regulations. Post-war, he formed the British Motor Racing Research Trust, which became British Racing Motors, the organisation behind the famous BRM Grand Prix project with its ambitious 1.5-litre supercharged V16

QUIET GIANTS

Sir Malcolm Campbell at his home, Povey Cross, with son Donald and the Railton-developed *Blue Bird*, now with Rolls-Royce power. The car is fitted with its transit tyres, from a lorry; the tyres for speed runs had barely an eighth of an inch thickness of rubber and canvas, to reduce balance problems. *Getty Images/ Topical Press Agency*

engine. Again there was outside backing, this time from Alfred Owen (Owen Organisation), Oliver Lucas (Lucas Electrics), Sir John Black (Standard Motor Company) and Lord Hives (Rolls-Royce). Hives later suggested that those consulted during the development of the V16 engine included Railton, with whom he had worked closely on various aspects of the Rolls-Royce R-type engine in record breaking, while input also came from Frank Whittle, who had used his knowledge of the supercharger impellors on the R-type engine in the development of the first jet engines, and from 'apprentice' Tony Rudd, who became pivotal in BRM's Formula 1 endeavours.

John Cobb

Shortly after Parry Thomas's arrival at Brooklands, John Rhodes Cobb, a wealthy man with a similar liking for big-engined racing cars, approached the Thomas Inventions Development Co. Ltd. and became an unofficial sponsor of its proprietor's quest to develop *Babs*, in exchange for being allowed to race her occasionally. Born into a wealthy family of fur brokers, Cobb had few interests outside business and cars, and many of his family and friends said they never really felt they knew him.

CRUSADER

The Cobb family had lived in and around Kingston-upon-Thames, Surrey, from at least 1700 and the family history explains the family names. Thomas Cobb married Sarah Hawtyn and they had a son, Timothy, who inherited large tracts of land in the Kingston area. He married Ann Rhodes in about 1730 and their son, born in 1748, was christened Timothy Rhodes Cobb. In turn, his first son, born in 1823, was called Rhodes Cobb. Rhodes married Lydia Davies, who were the grandparents of our John Rhodes Cobb. Rhodes was very influential locally and lived in considerable style at Surbiton Lodge, a fairly grand manor house set in four acres of grounds (the site of the manor is now occupied by St Raphael's Church).

Rhodes and his son, Rhodes Hawtyn Cobb, were both directors of the Falkland Islands Company, the latter serving as chairman from 1922 to 1929. The Cobb family had had interests in the Falkland Islands since 1867, when 22-year-old Frederick E. Cobb had arrived in Port Stanley and acquired the Falkland Islands Company. This firm supplied sheepskins to another family business, Anning Chadwick & Cobb, a fur broker that had been set up in the late 1700s by cousins Henry Anning, William Chadwick and Arthur R. Cobb.

Rhodes Hawtyn Cobb married Florence Goad and they lived on Croydon Road, Beddington, Surrey. John Rhodes Cobb was the fourth of their five children and was born on 2nd December 1899, just before the growing family moved to the Cobb family home — Grove House, Arran Way, Esher, Surrey. John's siblings were Rhodes Stanley (born in 1890), Olive (1893), Gerard Rhodes (1895) and Eileen Lucy (1902).

The cover of a Falkland Islands Company annual report, displaying the names of both John Cobb's father, Rhodes, and an older forebear, Frederick, who had acquired the company 39 years earlier. Although the Cobbs more or less owned the company, the family trait of 'blending in' is evident. *Cobb archive*

This is the Cobb family home, The Grove (later Grove House) in Esher, Surrey, depicted upon its completion in 1792. It fell into disuse after John Cobb's death and was eventually demolished. *Courtesy of Diana Sweeney*

Cobb's introduction to the motor car came during childhood. He learned the basics of how to drive with his oldest brother, Rhodes, by experimenting illicitly with the family doctor's De Dion, rolling it along the drive at Grove House. The first time they did this, they realised they could not push the car back to where the doctor had parked it, but they succeeded in starting the engine and reversed it back. This led the pair to welcome the doctor's further visits so that they could try out the forward gears as well. After some proper driving experience with their father's Panhard and a two-stroke Valveless, the brothers got their hands on a wrecked Minerva, fashioned some 'Heath Robinson' repairs and made it run well enough to carry them to Leatherhead, six miles away, before it expired for good. Although John's education left limited time for indulging in his growing passion for all things mechanical, he and Rhodes also took to motorcycling along with sister Eileen, travelling far and wide, without any need to travel quickly, merely to enjoy the adventure.

As a child Cobb was also a keen cyclist and at the age of eight he worked out cycling courses near his home so that he could time himself, aiming to improve. Eventually these routes led him to nearby Brooklands, where he became a regular visitor. He watched the racing there both closely and close-up, often being asked to stand further from the edge of the track. He began to make notes about the faster drivers and how their driving differed from those in the slower cars. Lines, braking points, overtaking areas and techniques were all written down, accompanied by very skilful sketches, including suspension layouts and body styles. He came to know the Brooklands track probably better than anyone who ever raced there.

His brother Gerard later said, "That book went everywhere with him. One way it was cars, gears, driving equipment and ideas, cleverly turned upside down, it was fur prices, sizes, requirements, uses and suppliers. It was in his pocket when he died..."[3]

At the age of 16, a severe problem affected Cobb's spine and had such a debilitating effect that when he turned 18, ready to enlist as a soldier in the First World War, he was rejected as unfit. This affected him greatly — especially when he later learned that two other speed heroes, Henry Segrave and Malcolm Campbell, had served in the war. He feared that those about him might consider him a failure, that he was incapable of success, and for the rest of his life he never once developed views of himself other than that of "a normal, everyday chap, no more important than anyone else".[4] He may have thought that, but he always admired those whom he felt were a 'cut above the rest' — and one such was Reid Railton.

In the years after the First World War, Cobb returned to Brooklands and made a point of meeting as many of the drivers as he could, drawn to those who raced the really big cars. Gerard Cobb recalled that in the early 1920s his brother approached Malcolm Campbell in order to get his autograph. In a brief conversation, Cobb told Campbell of his wish to be racing soon himself, and Campbell organised a ride for him with Ernest Eldridge as passenger

During the Second World War, John Cobb at first volunteered to drive buses in the streets of London, echoing the deeds of his father, who had volunteered as an ambulance driver during the Great War. *Ministry of Defence Records Department*

Seated in Parry Thomas's *Babs*, John Cobb awaits the start of a race at the BARC Whitsun Meeting at Brooklands in May 1926. Cobb's choice of attire at this time — entirely in black and always with a tie — came about because of the amount of oil *Babs*'s Liberty engine threw into the cockpit. *Motorsport Images/LAT*

in the massive Fiat, *Mephistopheles*, in which Eldridge took the land speed record at 146.01mph in July 1924 at Arpajon in France. Eldridge, no mean driver himself, was taken aback by Cobb's insight, his probing questions and suggestions, stating later, "This man will make a better driver than I."[5]

Cobb's racing baptism came at the 1925 BARC Summer Meeting at Brooklands. He organised a drive in Richard Warde's ex-John Duff FIAT *Tipo S.61*, which dated from 1910 and had a huge 10-litre four-cylinder engine with chain drive. Despite being a difficult car in which to make one's début, Cobb managed to finish in third place. Another Brooklands meeting staged by the West Essex Motor Club featured a challenge race between Parry Thomas in his *Leyland-Thomas* and Eldridge in *Mephistopheles*. Amidst much gossip that this race was beyond the capabilities of both the cars and their tyres, many other drivers went to the bar until it was over, refusing to watch. Parry Thomas finally won, after both cars lost tyres in the final run to the flag. Standing beside the track was Cobb, keen to see how the big FIAT would perform. He is said to have turned to Warde and commented, "To put that Leyland that close for the entire race, and take the win over a monster of a car, shows some

CRUSADER

Competitors in the British Empire Trophy race at Brooklands in 1932 wait for A.V. 'Ebby' Ebblewhite to flag them off. John Cobb is nearest in his Delage, with Tim Birkin's famous single-seat Bentley alongside and George Eyston's Panhard next in line. Eyston was a rival of Cobb's both on track and in record breaking, but they remained firm friends right until the end — when Eyston was *Crusader* project manager. *Motorsport Images/LAT*

impressive engineering."[6] After his own race in Warde's *Tipo* S.61, in which he scored his first win, Cobb sought out Parry Thomas in the paddock in order to discuss his own plans.

In February 1928, nearly a year after Parry Thomas's death, Cobb bought himself a former land speed record holder. This was the 10.5-litre V12 Delage that had first appeared at the Gaillon hillclimb in 1923 and had continued to be successful in the hands of crack driver René Thomas, who drove the car to victory in countless events all over France. In July 1924, Thomas took the Delage to Arpajon and achieved a record speed of 143.3mph, only for Eldridge in *Mephistopheles* to beat it six days later. Seeing no use in tuning the car for further record attempts, Delage eventually retired the V12 from competition and offered it for sale. Cobb travelled to Paris to buy the car for £350, together with a spare engine and a host of extra parts, including a choice of seven rear-axle ratios, and had it shipped to London.

By this time Parry Thomas's legacy lay in Thomson & Taylor, and Cobb

John Cobb's new contender, the *Napier-Railton*, under construction at Thomson & Taylor, with Cobb and Ken Taylor looking on at right. The unmistakable shape of the 'Broad Arrow' 12-cylinder Napier Lion engine sits in what at first sight appears to be a basic chassis, but Reid Railton had included many features that made the car supremely successful, both on track and in record breaking. *Courtesy of Charles Taylor*

followed his earlier instincts in having his Delage garaged and tuned there, under Reid Railton's guidance. The Delage became a remarkably consistent performer, easily the equal of Tim Birkin's Bentley in 1932, the year Cobb carried off the dramatic 100-mile BRDC British Empire Trophy Race against George Eyston at an average speed of 126.4mph. In total, the Delage gave Cobb nine race wins, eleven second places and four thirds at Brooklands, as well as ten Class A records (for cars over 8 litres) and a record lap at 133.88mph. It seemed that Cobb now had the bit between his teeth.

With the Delage showing its age, Cobb sold it at the end of the 1932 season to Oliver Bertram, a young barrister who was beginning to make his name as a racing driver, and eventually the car passed to the Junior Racing Drivers' Club for use by members for tuition. To replace the Delage, Cobb commissioned the construction of his own car, designed by Railton and built by Thomson & Taylor. This car, which was to become known as the *Napier-Railton*, was as innovative as it was large. Powered by a 24-litre Napier aero engine, it was used by Cobb to set numerous circuit and distance records in Britain, France and the United States, and finally left the Outer Circuit record at Brooklands at an unbeaten average speed of 143.44mph.

Although Cobb is forever associated with driving cars of considerable size, both in dimensions and engine capacity, he was also successful at Brooklands in cars of more modest size. Driving Jack Barclay's 1922 3-litre TT Vauxhall during 1927, he won a '100mph Short Handicap' and finished second in a '100mph Long Handicap' and the prestigious Gold Star race.

With today's knowledge, it is difficult to say how challenging it was to race on the Brooklands oval: it certainly was not simply a case of applying full throttle and holding on, although bravery certainly played a part. Railton himself was once drawn into passing comment on record-breaking as an activity, stating, "You go out, you put your foot down and you get away with it, or you go out, you stop it halfway and you ask for a pot."[7]

In 1929, Cobb appeared in a true road race, the Ulster TT at the Ards circuit. To learn the track, he and his sister rode round on their motorcycles, Eileen remarking, "Left AND right turns John, and you have to slow down every so often!"[8] Cobb was entered in one of Victor Riley's 'Brooklands Model Nine' cars but, although he practised at every opportunity, he only lasted until the sixth lap before hitting the bank at Comber and retiring on the spot. He continued to hone his road-racing skills but undoubtedly preferred the discipline of long-distance competition at high speeds.

Although John Cobb had already been working at Anning Chadwick & Cobb, his life changed in 1930 following the death of his father, which brought new responsibilities as well as wealth. Already quite comfortably off, after his father's death he became much richer as the monetary estate was over £235,000 (equivalent to about £15 million today). With his mother and siblings, he became involved in the supervision of the family-owned farm that overlooked Sandown Park racecourse as well as playing a bigger part in the fur-broking business. From 31st December 1929, the title of the family firm changed from Anning Chadwick & Cobb to Anning Chadwick & Kiver, the Cobbs having decided that there was no longer any need for their name to appear in the company's title, even though they occupied most of the directors' seats.

Upon the outbreak of the Second World War, 39-year-old Cobb enlisted in the Royal Air Force, having qualified for his pilot's licence at Brooklands on 10th April 1924. To his dismay, however, his physical size made it difficult for him to get in and out of the cockpits of the fighter planes he longed to

QUIET GIANTS

The *Napier-Railton* was revealed to the press at Brooklands, photographed along with all those who built her. It was an occasion when neither Cobb nor Railton could hide from the camera. *Courtesy of Charles Taylor*

The *Napier-Railton* in its natural environment, flat out on the banking at Brooklands during the Easter Meeting of 1937. Cobb had visited the track from his mid-teens and knew it intimately. The concrete banking had been laid on sand and had sunk in places, but at this point, where the Hennebique Bridge crosses the River Wey, there was a more solid foundation. Having studied the area closely, Cobb knew the exact spot to cross and claimed that he was never intimidated about taking flight there. *Motorsport Images/LAT*

CRUSADER

Like so many others with a pilot's licence, Cobb yearned to fly fighters during the Second World War but a back condition precluded active service. He managed to convince the selection board to allow him to become a ferry pilot, a role in which he repeatedly volunteered to fly delivery missions as close as he could to the battlefronts.
Cobb archive

fly. Taking inspiration from his father, who had volunteered as an ambulance driver for the Red Cross during the First World War, Cobb instead drove buses through blitz-torn London. He then applied to join the Air Transport Auxiliary and during the last two years of the war he flew almost every day in a wide variety of aircraft, attaining the rank of Group Captain. One man who flew with him regularly, Captain Harold Peters, said of him: "He was the finest type of Englishman, self-effacing, extremely modest, completely imperturbable, a kindly man."[9]

If there was one person to whom Cobb would always defer, it was his mother Florence, whose opinion he valued above all others. Cobb's brother Gerard believed this maternal bond, together with shyness around women, explained why John took so long to marry. In 1947, just weeks before leaving for his last record attempt at Bonneville, he married Elizabeth Mitchell-Smith, of whom Florence approved greatly. Sadly, the union lasted only 12 months as Elizabeth succumbed to Bright's disease (chronic inflammation of the kidneys). Both Gerard and Eileen Cobb thought their brother knew of the ailment even before the marriage, but it did nothing to reduce the devastation he felt at Elizabeth's loss, while all the time refusing to share his grief with anyone, including those close to him. His mother wrote to his sister Olive,

"John is seemingly coping well, perhaps a little too well. I have been unable to get him to talk about Betty; he just carries on, hiding in his business and cars."[10]

In 1950, mutual friends introduced Cobb to Vera Henderson, and noticed a return to his old self. Born Vera Victoria Siddle, she had married Deryck Farquharson Henderson in 1942, but had divorced shortly after the end of the war. She recalled in 2005 how her divorce changed her outlook on life, including a wish to change her name from Vera: "It was an old-sounding name, so I took up Vicki, and indulged my interest of interior decorating until I was able to earn a comfortable living — quite novel for a single girl."[11] Her demeanour and manner were much the same as Cobb's, and the two married quite soon after meeting. She summed up her husband thus:

> John was by no means flashy, like a lot of his racing friends, not that he spent a lot of time with them, he was happier alone in his cars, or at home with a newspaper and a whisky and soda. John wasn't a 'social' animal. He was *'sociable'*, but not one for clubs, restaurants and groups, though he was relaxed in the company of those he knew, and he had a very special sense of humour with his friends.
>
> John was reserved, yes, a far better word than shy, he never did anything for publicity, or anything that might be seen as simply puffing himself up, that wasn't John at all. I would mention to him that I thought that perhaps he should stand near the front, or speak to someone, and he'd raise his eyebrows and say, 'No, no, I'm not Sir Malcolm you know…' Sammy Davis [Sports Editor of *The Autocar*, racing driver and Cobb's biographer] told me he used to ask John out to lunch to see what he was up to, they'd have a wonderful meal, with lots of chat, but at the end of it he was none the wiser![12]

Soon after the end of the war, Cobb had bought a house called Cullford in Coombe Park, Kingston Hill, close to Richmond Park and some six miles from his childhood home at Grove House. It was there that he and his second wife Vicki shared their brief, exciting life together.

After Cobb's death, Vicki became what one friend described as a "glorious woman".[13] However, she stated quite bluntly that when she met John she was quite meek and shy, but her life with him taught her strength of character, application and determination, and that her husband had nothing but a positive effect on those he met, especially children, with whom he seemed to have a special affinity.

Noted author and historian Doug Nye holds Cobb in very high esteem and recalled: "I remember Sammy Davis saying how difficult it was to get more than a polite greeting out of Cobb in the Brooklands days, and yet during wartime the great man put in that appearance in 'Target for Tonight' [a 1941 British documentary film about the crew of a Wellington bomber taking part in a bombing mission over Nazi Germany] and was apparently very proud of the fact that he did so. Cyril Posthumus [a respected journalist with a special interest in the land speed record] told me how much the generally detestable Malcolm Campbell hated Cobb attending any of his many presentations and talks. Campbell would be articulate and fluent until he noticed Cobb sitting in the audience, simply staring at him in total silence. And once fixed by that gaze, Campbell would usually go completely to pieces, stumble over his speech and get generally flustered and uncomfortable."[14]

Andrew Whyte, who became press officer at Jaguar and later a historian and author specialising in the marque, grew up around Inverness and his father was manager of the Caledonian Canal, which connects the west and east coasts of Scotland via Loch Ness. Constance Whyte, Andrew's mother, wrote one of the earliest books, if not *the* first, on the Loch Ness Monster entitled *More than a Legend*. She and her husband introduced their son, a shy schoolboy at the time, to Cobb on the quayside at Temple Pier during the early runs of *Crusader*, and Andrew recalled how he was overwhelmed at meeting the fastest man on earth: "He was absolutely enormous, a giant of a man, and he was terribly famous."[15] But as Cobb leaned down to shake his hand, Whyte said he had a "real impression of warmth and friendly kindness", and that he was able to ask questions that Cobb happily spent time answering.

It seemed that Cobb was more at ease talking with children than with adults, apart from his really close circle of friends and confidants such as Reid Railton. Yet Cobb's young niece, 'Bunny' Sweeney, recalled, "I spent many happy hours at his home, playing in his office while he worked. He'd walk over and join in with what I was doing, laugh and then the telephone would ring, and he'd go back to his desk, smiling over the top of it at me and waving, and yet I hardly really knew him."[16]

Cobb's sister Eileen had married Dudley Holloway in the 1920s, and both were fond of holding dinner parties, to which John would be invited on occasion. Dudley recalled: "You would welcome John and others into the

Together at Bonneville. Reid Railton and John Cobb on the salt after successfully raising the world land speed record, with both hoping to avoid using the press microphone in front of them. *Courtesy of Charles Taylor*

house, all sit around a table, eat, drink and make merry. Conversation would flow, and at the end of the evening as everyone took their coats and left you would suddenly remember John had been there! He'd have been involved, he had added to the evening, but by no means had he taken over, or spoken of his exploits, but added just enough to be a good guest, but you could easily forget he was even there."[17]

As Doug Nye observed, "To my mind Cobb was never really up to much as a racing driver *per se*, his road racing results aren't that impressive, that's for sure, but he was certainly very comfortable around the Outer Circuit and in his record-breaking exploits. Like many shy or enclosed people, he evidently quite liked the opportunity to perform before a large crowd, as long as he didn't have to address them or look them squarely in the eye."[18]

Cobb's nephew, John Hawtyn Cobb, was extremely proud to have the same name as his uncle, and spent as much time as he could with his namesake from an early age, even taking his first driving lessons from the land speed record holder: "He had so much patience, I think that is why children took to him the way they did. He refused to be rushed and was considerate and considering, a very deep thinker. With children, he'd just listen intently to every word, until he had a little group around him, and he wouldn't leave until they all had their say. He was the same with business and cars, everything was thought through. Everything had a purpose and a right time and he wouldn't be swayed from it, you couldn't pressure him, he wouldn't do anything he felt uneasy about, but he had a disarming way of letting you know if he wasn't going to do it!"[19]

Cobb put his engineering knowledge to good use too. He was an exceptional test driver, and his relationship with Reid Railton, Ken Thomson and Ken Taylor was based on mutual respect, both as people and as engineers. Cobb was capable of relating precisely the behaviour of his vehicles at any given time or speed, but would follow to the letter any instructions issued to him.

He took a deep interest in his machinery, questioning Railton to explain anything he did not fully grasp, although that was rare. When he arrived at Bonneville for his first land speed record attempt with his *Railton Special*, in 1938, he had already spent many hours at Fort Dunlop in Birmingham watching high-speed tests of the car's tyres, and as such had learned how to avoid wheelspin while still accelerating quickly enough to make full use of the available course. He had the ability to sense how his tyres were wearing through, both as a set and individually, all from a cockpit mounted at the very front of the car — and rarely were his assessments wrong.

Cobb was a typically stoic Englishman who was also an accomplished engineer in his own right, capable of formulating solutions to problems but also happy to surrender himself totally to those whom he considered more knowledgeable.

Reid Railton and John Cobb were men from the same mould. Both were quiet, unassuming and more than happy for others to take centre stage while they melted into the background. Because of their similarities, they became firm friends very quickly. While Parry Thomas was still alive, indeed, they formed a close-knit trio and were often referred to around the clubhouse at Brooklands as 'The Three Musketeers'. Even though over the following years Railton and Cobb pioneered many revolutionary ideas, Cobb had implicit faith in what he described simply as "Reid's genius".

Cobb was much like his friend and test pilot Geoffrey Tyson, who once said of the quiet giants behind the *Crusader* project: "They are putting themselves on the line, and if it works, you have succeeded, but if it doesn't you might end up paying the ultimate price, though neither are anything less than careful and thoughtful in what they do, they simply recognise the risks."[20]

CHAPTER 2
FROM WHEELS TO WATER

As early as 1933, Sir Malcolm Campbell — who had been knighted in 1931 — had confided in Reid Railton that once he had surpassed 300mph on land in his *Blue Bird*, he would turn his attention to the water speed record. At that time, the record was held by American Gar Wood at 124.86mph with his *Miss America X*, a single-step hydroplane powered by four supercharged Packard aero engines. Campbell achieved his landmark on 3rd September 1935 when he recorded 301.129mph at Bonneville Salt Flats in Utah, becoming the first man to break the outright land speed record at that hallowed venue. With that, Campbell did indeed embark on his new endeavour and asked for help from Railton. Both men were new to the world of high-speed boats.

Working to Campbell's instructions, Railton had to utilise the *Blue Bird* car's Rolls-Royce R-type engine and design a new drive system for it, a task that was not made any easier by having to package it all in a hull designed by someone else — and while contriving to be as sparing as possible with Campbell's money. The project fascinated Railton as he had already become aware of the marked differences between aerodynamics and hydrodynamics.

The first *Blue Bird* boat, *K3*, was fairly conventional for the time. Its designer was Fred Cooper, who freely admitted that Railton greatly influenced the exercise, bringing in ideas that a marine architect would never have considered. Cooper's previous work had included producing the drawings for the craft that became *Miss England*, in which Sir Henry Segrave first learned high-speed boat control, and then *Miss England II*, in which Segrave took the record on 13th June 1930 only to lose his life immediately after doing so. That day, on Windermere in England's Lake District, Segrave had just set a new two-way record of 98.76mph when he turned to make another run, only for *Miss England II* to capsize at speed, killing him and engineer Victor Halliwell. The second engineer (there was one for each engine), Michael Wilcocks, survived with bruising but suffered for years afterwards with what is now known as Post-Traumatic Stress Disorder.

Built by Saunders-Roe, Campbell's first boat was designed solely to break the water speed record and was as compact as it could possibly be. For balance and general weight distribution, everything was mounted as close as possible to the craft's centre line, so there would be room only for the helmsman. Cooper knew that structural strength would be critical at the speeds needed to break the record but deferred to Railton on how that strength could be achieved. The scheme they arrived at had the centrally mounted engine bolted to longitudinal frames of Canadian rock elm, boxed with diagonally jointed three-ply. This central armature was completed by the forward-mounted gearbox bolted to the frames. Railton created a V-drive by turning the engine around so the drive was taken forward from the engine, through a dog clutch and to the gearbox, where the rotation was geared up by a ratio of three and reversed back to the stern-mounted propeller.

This clever packaging allowed for a short, small hull, while keeping the weight in the right place and the propeller where it needed to be, at the stern without the need for a complex vertical drive. It also allowed Railton to keep the stresses and torsional forces where he wanted them, and meant that Cooper could design a hull with the necessary buoyancy and planing surface for the weight and dimensions, and with little concern for the mechanicals of it all. Starting with the central armature, Cooper's design attached frames of differing profiles, bow to stern, creating the boat's shape, like a massive scale model. These frames were cut from single sheets of seven-ply, to maintain their strength. Cooper wanted these frames attached to the central runners by wooden frames, but as he later said: "I am a designer of boats, fairly traditional boats. Railton suggested using triangular Duralumin brackets with the support members for the deck fixed to them, very 'aircraft' in thinking. It was stronger and it was lighter, and at a stroke reduced the internal structure by 25%. It was a fine marriage in thinking."[21]

The hull was completed by sheathing this frame structure in a mixture of six-ply, five-ply, and cross-laid mahogany planking. When complete, *K3* had an all-up weight of just 2½ tons, or 4,945lb to be exact. With an engine

FROM WHEELS TO WATER

The press launch at Brooklands for the 1935 incarnation of Sir Malcolm Campbell's *Blue Bird*. The car still used all of the original chassis and many other components from the first purpose-built land speed car of 1927. Reid Railton designed the all-enveloping, aerodynamic bodywork on paper, before presenting Vickers Ltd with a wooden model to test in its wind tunnel. Only one small modification was needed to see Campbell become the first man to surpass 300mph on land. *Getty Images/Fox Photos*

weighing 1,635lb and the gearbox and drive shafts additionally taking up just over a ton, the bare hull weighed less than 1½ tons. *K3* measured 22ft 3in at the waterline, with a beam of 9ft 6in. Because of her wide, gently curved under-surface, her draught was a mere 1ft 9in at rest.

Although an unstable craft, because when planing she was supported on two central points (the very centre of the forward step and the stern), she lifted the record twice in 1937, and once more in 1938. However, when Campbell took *K3* to Lake Hallwyl in Switzerland for her last record, 130.91mph set on 17th September 1938, it became obvious that she was going as fast as she could. Both Campbell and Peter Du Cane, the Vosper designer whom Campbell had consulted along with several other boatbuilders, found her unstable and difficult to hold.

Curious by nature, Railton had been investigating a different planing principle and had obtained drawings that both Campbell and Du Cane found intriguing. But if Campbell was to increase the record with a new boat, he

CRUSADER

Fred Cooper during testing of the hull shape that was to become *Miss England II*, the model being pulled by a frame attached to the accompanying boat. Later, Railton would suggest using small rocket motors. *Author's collection*

Sir Malcolm Campbell's *Blue Bird K3* in 1937 on Lake Maggiore during early trials. The cooling ducts for the engine got ripped off and the gearbox bearing overheated. When the record was finally taken, the aerodynamic tail had gone, replaced by myriad water pipes, and the horn from Campbell's Rolls-Royce had been pressed into service as a cooling scope for the gearbox. For the following year, the water-cooled exhausts had also gone. *Author's collection*

would insist on using the drivetrain from the old one to save money, as usual.

Railton arranged to borrow a small motorboat, *Blue Ace*, based on the new design. Both Campbell and Du Cane tried it out while Railton made copious notes. That evening, the three men were joined over dinner by Leo Villa, Campbell's long-time mechanic, for discussion about the craft's performance. Railton's notes included a photograph of *K3* on which, according to Villa, he had made annotations: "Railton had altered [the picture] with a red pen to see how the new principle could be 'wrapped around' the already existing drive train. The 'Old Man' got quite excited and Du Cane remarked, 'Yes, we could build that.' Railton just rocked back in his chair and rubbed his hands."[22]

Meanwhile, Railton continued work for speed on land. Even as Railton and Reg Beauchamp had looked over the drawings for Campbell's *K3*, mere months after Campbell recorded his 300mph land speed record, they had been visited by John Cobb. As Beauchamp recalled, "I was standing beside Reid, looking over his shoulder as he sketched out how tightly he thought we could package the new hull when Mr Cobb came in, and a wry smile came to his face. 'Looking after the purse strings again chap?' To which all Reid could do was huff. He [Cobb] leant back against a drawing cabinet, hands deep in the pockets of this long coat he had, and looked across the desk and said, 'Look, if you had a reasonable amount to play with, what would you need, to come up with something to raise the land speed record?' I remember Reid leant back in his chair, he always had it so far from the desk and leant forward from it, and cupped his hands behind his head, and sort of looked at Mr Cobb from under his eyebrows, and said, 'A clean sheet of paper would come as a breath of fresh air, if you're serious?' I think we had four or five sketches done before they left for lunch."[23]

Alongside John Cobb's new land speed record project, Railton continued to advise Sir Malcolm Campbell in a further association with Fred Cooper, who was now working freelance, to design a new *Blue Bird* boat, *K4*, which Campbell commissioned in the autumn of 1938. Based on drawings Railton had obtained from Adolph E. Apel, the new craft had what was known as a 'Ventnor' hull, of the type that Railton had shown Campbell at Lake Hallwyl. The Ventnor hull was also known as a 'Three Pointer', invented and patented by Apel and the Ventnor Boat Works, Atlantic City, New Jersey, USA.

The Ventnor hull had come about a few years earlier, when the company was asked to design a 'suicide boat' for the Chinese government, the brief

CRUSADER

Blue Bird K4 under construction at Vosper. After working with Fred Cooper on *Blue Bird K3*, Reid Railton suggested the increased use of Duralumin in the new boat, as is clear in this photo. Cooper had left by this time, due to conflicts in the design process. *Author's collection*

being a speed of 50–60mph with the ability to carry a 500lb bomb. Initially problems with stability appeared difficult to overcome within the given criteria, but Apel suggested the addition of a pair of Ventnor water skis either side of the hull. Not only did this cure the stability issues, but Apel soon realised that once the boat was up on these skis, drag on the underside of the hull was greatly reduced and speeds shot up. Ski-like forms were therefore permanently built into the front of the hull as dedicated planing surfaces, designed to lift the hull out of the water. When matched to a concave bottom to the hull, the entire craft ran on these two forward surfaces and a third positioned centrally on the stern — hence the name 'Three Pointer'. Apel had calculated various formulae for this system, allowing for stability and speed, as he found that if the bow lifted beyond its safety margin, aerodynamic forces (rather than hydrodynamic forces from the water) increased on the forward area of the hull and could cause a Ventnor hull to 'kite' upwards — or, as it is known today, suffer a 'blow over'. This type of backward somersault and crash would occur with two subsequent water speed record craft, Stanley Sayres's *Slo-mo-shun IV* (1950 and 1959) and Donald Campbell's *Bluebird K7* (1967), although there were also other influences in Campbell's accident. Incidentally, it should be noted that Donald, unlike his father, used one word for *Bluebird*.

FROM WHEELS TO WATER

Returning to *Blue Ace*, this boat had been built from spruce-covered plywood and was powered by a 4-litre 175hp Lycoming engine. Piloting her, Harold Notley had unofficially reached speeds of over 100mph, not too far short of the 138mph that Sir Malcolm Campbell had set as his new target, but with far less power. Notley also established several world water speed and endurance records during 1937–38 on a measured course at Poole, Dorset, not far from the Portsmouth-based Vosper — although it seems that the whole Ventnor concept went unnoticed by the company.

It was during this period that Railton first met Douglas Van Patten, a marine architect cut from the same cloth. Born in Detroit, Van Patten served an extensive apprenticeship during the late 1920s with George W. Crouch, a notable naval architect, before going on to graduate in mechanical engineering from Columbia University in 1933 followed by post-graduate work in hydrodynamics. He became a consulting mechanical engineer for William O. Smith & Associates and then in 1935 joined the Eddy Marine Corporation of Bay City, Michigan, as its chief naval architect. While there he designed the Eddy AquaFlow, a 19ft runabout powered by a Ford V8. This craft was quite similar in layout to *Blue Bird K3* and included a V-drive.

In 1937, Van Patten moved again, this time to Greavette Boats, Ltd. of Gravenhurst, Ontario, Canada, again as chief naval architect. He was responsible for the design of the 'Streamliner' and 'Sheerliner' models, which continued in production for 38 and 29 years respectively. He also invented a machine capable of carving round planking from standard stock, making production of these radical craft possible. By 1939, Van Patten was chief designer at Minett-Shields, of nearby Bracebridge Falls, Ontario, where he stayed for five years. Even in these early days, Van Patten's thinking, when it came to a boat's shape below the waterline, was quite different from 'standard' practice and it was perhaps inevitable that eventually his path should cross with that of Reid Railton.

While Railton was negotiating with Apel to borrow *Blue Ace*, Stanley Sayres was considering purchase of a Ventnor boat, but it was too expensive for him and instead he went looking for a more cost-effective way of laying his hands on one of the hulls. Finally, in 1942, Sayres bought *TOPS III*, a Ventnor-designed 225 Cubic Inch Class hydroplane, from Jack 'Pop' Cooper of St Louis, Missouri. During transit to its new owner, one of the sponsons on *TOPS III* became damaged and had to be replaced, so Sayres asked Seattle-based Tudor Owen 'Ted' Jones to help realign the new one. Once repaired,

CRUSADER

TOPS III became *Slo-mo-shun II*. At about this time, Jones told Sayres that he had the working design for an extremely fast boat and was seeking a backer, and this craft duly became *Slo-mo-shun IV*.

Ted Jones had already developed quite a reputation as a designer of high-speed boats and had shared data with Douglas Van Patten. At some stage during the early 1940s, exchanges began between Sayres, Jones, Van Patten and Railton, which explains how, in the years to come, Railton was able to go to Sayres and simply request and receive data on the prop-riding principle used for the 'extremely fast boat' that Jones had suggested to Sayres in 1942.

During the Second World War and for a period afterwards, *Blue Ace* vanished, but she was subsequently discovered by none other than Donald Campbell in the Lake District, where she had been a tender to Sir Malcolm Campbell's *Blue Bird K4* jet conversion in 1948. It was apt, therefore, that the rebuilt *Blue Ace* should be used by Donald as a fast tender for his own record attempts. In the 1970s, *Blue Ace* was bequeathed to Britain's National Maritime Museum.

Returning to the genesis of *Blue Bird K4*, eventually Fred Cooper took a set of his drawings for the craft to Brooklands, and Railton's drawing office at Thomson & Taylor. Both men quickly realised that, from the waterline down, they were pretty much tied to using Apel's formulae, but in taking into account that their design had the engine in the stern, they ended up with a longer, wider and slightly flattened version of *Blue Ace*. As was always the case, Railton's office became a meeting room, where a few options were offered by those present, who numbered Ken Thomson and Reg Beauchamp besides Railton and Cooper. Beauchamp remembered the meeting as being, "Humorous, but constructive, usually with a cup of tea or four! I do recall Freddie [Cooper] was quite concerned about one Telex from Apel, about the 'kiteing' risk at the bow, and he suggested that before the final outline was taken down to Vosper, who had agreed to build it, that we might add some form of artificial turbulence generator at the bow. He was quite put out when Reid told him later that [Peter] Du Cane had said it was 'needless'. He decided there and then not to bother travelling up from the Isle of Wight any further, as he thought there was, 'Little left to actually design.' It was a shame, I rather liked Freddie."[24]

Although 'Freddie' walked away from the project, he was later proved right regarding the design of *K4*. Vosper turned the 'concept' drawings into engineering drawings that the company's workshops could use during construction. Models were tested both in the towing tank at Haslar (the Admiralty Experiment Works

Sir Malcolm Campbell at the press launch of *Blue Bird K4* at Adams Transport, New Malden. Clearly visible is the bulbous nose suggested by Fred Cooper, to cancel out aerodynamic lift at the bow, something Vosper thought unnecessary and became one of the contentious issues that led to Cooper's departure. *Author's collection*

at Gosport, Hampshire) and the wind tunnel at Vickers (at Brooklands), where one model showed a tendency to back-flip at scale speeds approaching 140mph. Consequently, the design was changed to incorporate a slightly bulbous nose, to generate a small amount of downforce at the bow.

Although Cooper left the record-breaking arena, in another of the coincidences that seem to inter-connect all those who were involved, he later became involved with Donald Campbell and his Jetstar project, a mass-produced waterjet-propelled leisure craft.

Campbell's *Blue Bird K4* can best be described as 'evolution' rather than 'revolution'. It took the basic concept of *K3* — central armature, two longitudinal rails, rearward engine and forward gearbox — but differed in its increased use of lightweight metal alloys in both the frames and bodywork, as suggested by Railton's friend Dr Ewan Corlett, chief engineer at British Aluminium, whom Railton had met when both men were working on one incarnation of Campbell's *Blue Bird* car.

Even as early as May 1938, Campbell was trying to replace his ageing Rolls-Royce R-type engine with the then-new Merlin. In a letter dated 21st July 1938, A.F. Sidgreaves of Rolls-Royce went as far as to say, "We should be very pleased to supply you, but in view of the pressure at present being put on us for the supply of all the engines we can produce, it will be necessary for you to obtain Air Ministry permission for us to supply you."[25] With Britain drifting towards war, of course, the supply of an engine was unlikely.

A free engine was one thing, yet Campbell still felt the record could be raised more cheaply, as shown by this letter to Peter Du Cane, dated 19th June 1939:

> I received your account this morning which has shaken me to the core as I had no idea that the cost would be anything like as high as this.
>
> My last boat cost £900 complete, including engine installation etc., even down to the propellers, rudder, stern bracket etc. It is true that they [Saunders-Roe] lost money and I ended up giving them an additional £250.
>
> You can appreciate, therefore, that I did not realise that the hull itself would cost close on £2,000, seeing that I have to bear the expense of the engine installation and that so many parts from the old boat are being used in the new one.
>
> In so far as the material is concerned, I see that timber cost over £220 and I believe that I could have obtained this, either free, or at any rate at a very large reduction had I realised that this would have been so expensive. You have also charged £8.10.2 for paint and varnish and here again I could have got this free. I really do object to paying for this item seeing that by merely writing a letter I could have obtained it for nothing.
>
> I would like you to go through this account again and see in what way you can reduce it as I am not the millionaire that either you or Reid [Railton] imagine. I hope therefore that you will be able to look into this matter.[26]

FROM WHEELS TO WATER

Blue Bird K4 at speed on Coniston, planing almost perfectly and showing her concave underside clear of the water so that she rides on the two outer points. *Author's collection*

When Du Cane brought this letter to Railton's attention, Reid raised his eyebrows, and said: "I wouldn't mind, but I've re-used so much, even from the cars, to save money, and as yet, I'm not certain Malcolm has even paid me!"[27]

Regardless of the cost, it was only the onset of war that prevented *Blue Bird K4* from taking more than just a solitary record, 141.74mph on Coniston Water on 19th August 1939.

As the spectre of war grew over Europe, Reid Railton, at the suggestion of the British government, moved to the United States to work on a variety of secret war projects in collaboration with the Hall-Scott Motor Company in Berkeley, California. He later considered his input into the design of the quick-build 'Liberty Ships' to be the pinnacle of his work in that period. Even from America, though, Railton kept strong links with the Brooklands scene. With war came accelerated development in nearly all of the technical disciplines, along with better understanding of aerodynamics and hydrodynamics, construction techniques and methods, and, above all, engine development.

By June 1948, *Blue Bird K4* had changed considerably. Gone were the smooth lines, as had the piston engine that once lived inside them. Within days

CRUSADER

De Havilland's test house in March 1947: Frank Halford (left), Peter Du Cane and Sir Malcolm Campbell study the gauges as the Goblin jet engine allocated for *Blue Bird K4* is put through its paces on the test bed. *Getty Images/Hulton Archive*

of the war's cessation, Sir Malcolm Campbell had decided to make another record attempt and contacted both Vosper and Reid Railton. *Blue Bird K4* had been in safe storage during the war years and her engines kept at Thomson & Taylor. As Railton was by now dividing his life between the United States and, to a lesser degree, the United Kingdom, communication could be slow, but it allowed time for some calculations suggesting that the old Rolls-Royce R-type engine must be at its very limit and that Rolls-Royce's newer Merlin and Griffon engines would only increase the record marginally — if Rolls-Royce could be persuaded to supply either of them. Campbell, never one to miss an opportunity and a keen viewer of newsreels, instead became drawn to the new jet engine.

After a fairly hurried redesign of its upper deck, *K4* arrived at Coniston Water in August 1947 fitted with a De Havilland Goblin jet and a completely new look. She proved difficult to coax into planing, and when she did manage to do so she showed an alarming tendency to yaw uncontrollably off course.

Campbell complained bitterly to Du Cane and Vosper, who came up with a list of modifications that they thought would take three months to incorporate.

FROM WHEELS TO WATER

In 1947 Sir Malcolm Campbell returned to the Lake District, to launch the jet-powered version of *Blue Bird K4*, now nicknamed the 'Coniston Slipper'. Much was expected, but little delivered. *Author's collection*

Meanwhile a Telex to Campbell's 'MaloBell' address from Railton read: "Calcs done. Suggest blind alley, more via mail. RAR."[28] A letter from Railton soon followed and for Campbell it cannot have been good reading:

> Thank you for your kind gift, a taste of the old 'homeland' that is much appreciated. Its arrival prompted me to get this to you sooner rather than later, but I fear it won't be what you want to read.
>
> With the information you supplied from Du Cane, the problem becomes obvious. To alter a design, drawn to be as effective as possible with the dimensions set, and the engine and drive system we had to draw it around, to a new, and very different form of propulsion is, pure folly.
>
> To utilise the hull designed around the venerable Royce engine, for the radial flow jet (do you require the engine drawing to be returned?), would require far more that the simple expedient of installing it in the same place. The weight of the unit is being carried too high and hence also, is the thrust line. A few brief calculations would indicate that the

engine would need to be some three feet further forward, and over twenty inches lower, so exactly where your good self sits. Not ideal.

I also have concerns regarding the surfaces below the waterline. I understand from your comments, that the boat has been difficult to turn, but the size of the rudder that has been installed puzzles me greatly. Surely, as speeds increase, the rudder should become smaller? The drag must be considerable, and the resistance to inputs at speed huge?

I can see no quick and easy solution to this, although nothing is insurmountable, but wondered if you really have the time and resources to follow this particular path?

I have had correspondence with Du Cane on the matter, and feel that his model tests are doing you a disservice, as the thrust line does not seem to have been correctly reproduced, and I have suggested a better design of towing carriage.

I can confirm the dates of both my travel and meetings when next over, by which time I will have more to discuss with you.[29]

Of course, now that the problem had been presented, Railton was giving some thought to a solution, regardless of whether or not Campbell was going to dig down into his wallet and have a new contender built.

"Reid did a few rough outlines, and some basic calcs," recalled Reg Beauchamp, "basically a Ventnor hull, with the planing surfaces facing the other way, Reid's idea being, to have the prop' in-between the twin rear shoes, and the pilot up front above the single front shoe. But it meant a vertical drive box from the engine down to the prop', there wasn't the room for a 'V' drive, but with a jet..."[30]

This was mid-to-late 1948 — about the same time as John Cobb approached Railton with the idea of going after the water speed record himself.

Meanwhile, Vosper produced a further concept for *K4*, with smoother bodywork and a large underwater fin, but this work was never carried out and only the previously requested modifications were done. After a few secret runs in Poole harbour conducted at quite low speeds by Peter Du Cane, *K4* returned in its modified form to Coniston. The handling had not improved, and at speeds of around 100mph she porpoised violently and would yaw unpredictably off course. By now 63-year-old Campbell was showing his age and for the first time in his career he started to have doubts. Quite obviously battered from

FROM WHEELS TO WATER

Increased height, weight and power — and the tendency to porpoise — had a detrimental effect on the directional stability of the jet-powered *Blue Bird K4*, as can be seen here at Coniston as Sir Malcolm Campbell fights to control another skid. It was enough for the bruised and battered speed hero to consider giving up. *Author's collection*

the rough ride he was enduring, he started to arrive late for trials, and on one particularly misty morning he was over an hour late. Leo Villa:

> I had done all the prep' work and the engine boys were all set for the start up, but there was no sign of the skipper. He'd been getting rough-housed on every outing in the boat, and was very badly bruised all over. In truth, I think he'd had enough. Peter Du Cane came over to me, seemingly quite impatient to get things going. He wanted to witness another run, before he set off south again. As I told him about the state of the old man, he just cut me off, and said, 'Go and get him Villa, because if he doesn't drive it, I will.' I went to see the skipper, and he looked dreadful. Tired and grey, he silently waved me away. I said I'd let the press boys down, and mentioned how happy Du Cane would be, and told him [Campbell] what he'd said. 'Bugger that Leo, we'll see about that,' and off we went to the jetty. He looked daggers at Du Cane as he walked past him...[31]

CRUSADER

It was to no avail. *Blue Bird K4* was as unstable as ever, regardless of the modifications that Vosper had designed and installed.

She was returned to Vosper yet again, this time to have the fin at the very least produced and installed, although how more hydrodynamic drag could improve all-out speed is difficult to understand. The modifications remained incomplete when Sir Malcolm Campbell died on New Year's Eve, 1948. In the little black notebook that Railton always had tucked in his pocket was the note, "*K4*, too much weight to stern, bow needs pushing down."[32]

Wherever Donald Campbell's motivation came from, by August 1949 he had his father's *K4* returned to her prewar form and was stationed at the same boat shed on the shores of Coniston to carry out the same task as his father. He knew that his father had always sought out those who were the best in their fields, and before starting work on *K4*'s hull he had retained Leo Villa and had been in constant contact with Reid Railton. Happy to advise, Railton was unable to be more involved because of his trans-Atlantic commuting, and in any case by then Railton and Cobb were, as Beauchamp put it, "as thick as thieves, sketches and ideas flowing freely!"[33]

After some troublesome runs in the Lake District, Donald Campbell's camp was beaten to the record by Stanley Sayres in his *Slo-mo-shun IV* hydroplane, which achieved 160.323mph on 26th June 1950, breaking the existing record by a considerable margin. Railton capitalised on his earlier meetings to simply ask Sayres how it had been done, and the American was quite happy to share the technical details behind his success in the full knowledge that it would be used to try to take the record from him. To Railton, it was, as usual, simply a fascinating engineering solution, one he had so nearly solved himself.

As Donald Campbell had noticed on several record runs, the engine's water temperature had suddenly increased as the speeds had also jumped upwards. This turned out to be caused by the stern lifting, reducing the effectiveness of the cooling-water scoop because *K4* was effectively trying to 'prop ride'. In effect, the aft planing surface became the hub of the propeller, emulating the technique deliberately built into *Slo-mo-shun IV* by Ted Jones. Campbell and his team, far from disappointed, saw Sayres's success as a further challenge, and, with Railton conveniently based in America, dialogue via him between Sayres and the *Blue Bird* team led to another batch of modifications.

The engine was moved forward some 6ft and slightly lowered, which did indeed place it exactly where the cockpit had been. The engine's height was fixed by the angle needed for the drive shaft to work properly, especially as it

FROM WHEELS TO WATER

Blue Bird K4 at Coniston Water in 1949 during her third life, when returned to Rolls-Royce power by Donald Campbell. Working on engine number R39 are (from left) Sid Randall, Leo Villa, Harry Leach and Goffy Thwaites. *Leo Villa collection*

Stanley Sayres and *Slo-mo-shun IV*: Ted Jones's revolutionary design utilised his concept of prop riding, where the hub of the propeller becomes the stern planing point, lifting the prop shaft and one blade clean out of the water and reducing drag. It was a technique that Donald Campbell came across almost by accident, but would be implemented with the help of Reid Railton and his friendship with Jones. *Author's collection*

was now some 4ft shorter, putting an extra load on the gearbox and mount, as well as the dog-clutch assembly, because there was less 'give' in the new shaft. Ewan Corlett designed a new skeg (a sort of hinged rudder/stabilising fin that also has a hub for the propeller) and the now-crucial propeller while the team installed dual cockpits into the hull, directly above the planing shoes. Even though money was short, and the work carried out quite quickly so that Campbell could compete in the Oltranza Cup, on Lake Garda, Italy, in June 1951, the modifications worked very well indeed. Not only did they allow *Blue Bird K4* to compete and win in the races but also to get very close to breaking the outright water speed record.

Three months later, in September, *K4* was running again on Coniston Water when a massive mechanical failure put her beyond further use.

CHAPTER 3
FASTEST MAN ON EARTH

Following Sir Malcolm Campbell's retirement from record-breaking on land with *Blue Bird* after breaking the 300mph barrier on 3rd September 1935, the names of George Eyston and John Cobb came to the fore in the quest for the land speed record, fought out on the bleak expanses of Bonneville Salt Flats.

Eyston, a first-rate engineer and a descendant of Sir Thomas More, had been active in record-breaking for several years, and saw Campbell's withdrawal as his opportunity to add the outright land speed record to the numerous class records he already held. To this end he set about the design and construction of *Thunderbolt*, an enormous, twin-engined, eight-wheeled brute that weighed a colossal seven tons. Eyston's thinking was that this huge weight would simply counteract the wheelspin that had so dogged Campbell.

John Cobb had already visited the Bonneville Salt Flats with his 24-litre *Napier-Railton* in July 1935 and taken numerous long-distance speed records, then returned in September 1936 and gone faster still, topping 150mph for 24 hours. From there, it was a relatively straightforward step up to have Reid Railton design a larger car for him to have a crack at Campbell's land speed record. Railton relished the opportunity of starting a design from scratch after the constant, cost-cutting modification of Campbell's *Blue Bird*.

While Eyston had already worked out that he would require the most powerful engine available, the Rolls-Royce R-type, his plan was to use two of them. Cobb's problem was that all of the R-types were spoken for and instead he had to accept the gift of two 20-year-old Napier Lion engines. These came from Betty Carstairs, who had used them in her Harmsworth Trophy

CRUSADER

John Cobb's Railton Special on display outside Thomson & Taylor at Brooklands in 1938. The sleek aerodynamic wheel covers have yet to be fitted, and at this time the car was still fitted with a 'pop-up' air brake. National Motor Museum

motorboat *Estelle*. Whereas Eyston would have in the region of 2,800bhp at his disposal, Cobb would be restricted to about 2,400bhp — an interesting problem for Railton's active and inventive mind.

Less power meant that weight joined wheelspin as the enemy, so Railton disappeared into his office and threw himself into the calculations for project Q-5000, Thomson & Taylor's in-house code name for Cobb's land speed car. Having already formulated the idea that four-wheel drive was the solution to wheelspin, Railton designed an ingenious S-shaped beam chassis with the engines slung from each side at 10 degrees to the centre line. The rearward engine would power the front pair of wheels, the forward engine the rear pair.

To assess how best to envelop this running gear with bodywork, five scale models were produced and tested in the Vickers wind tunnel at Brooklands. As well as covering the mechanical components, Railton knew that it was crucial to present the smallest possible frontal area. The models were named: A) 'Three Wheeler', B) 'Cigar', C) 'Blue Bird with troughs', D) 'Long tail' and E) 'The Bun'. After the Vickers evaluation, the list shortened. Further tests made at the National Physical Laboratory in Teddington, Middlesex, narrowed the candidates down to just one: 'The Bun'. Although 'The Bun' did not have the smallest frontal area, Railton calculated that any increase in drag could be counteracted by making the car 'crab-tracked', with a narrower track at the rear than at the front. In fact the distances between the wheels were 5ft 6in at the front and 3ft 6in at the rear.

Railton also took considerable interest in the figures generated by the 'Three Wheeler' and confided in Reg Beauchamp that he was already thinking the layout might work for a high-speed boat, noticing that high-speed weaving

The clever layout of the two Napier Lion aero engines, slung either side of a spine chassis, is clearly evident in this photograph. Unlike Eyston's *Thunderbolt*, Railton decided not to mechanically link the two engines together other than at the throttle: the front engine drove the rear wheels, the rear engine the fronts. Railton maintained that he had decided on the shape of the body first, and this layout and the narrow rear axle was the solution to fitting it all in. *Courtesy of Charles Taylor*

CRUSADER

John Cobb, George Eyston and Reid Railton on the salt of Bonneville. Even though Cobb and Eyston were rivals for the land speed record, they were happy to share information and data — and the condition of the salt was crucial to both endeavours. *Courtesy of Charles Taylor*

when running with the single wheel to the rear was cancelled out when the single wheel was placed at the front.

Beauchamp was put in charge of producing the working drawings for Q-5000, assisted by Mr Hobbs. Thompson Motor Pressings Ltd was entrusted with manufacture of the chassis, which was built in four sections and hot-riveted together. The massive, all-enveloping body was produced by craftsman Bill Masters and one assistant. Working from simple plywood formers made from Hobbs's drawings, they built a one-piece body that could be lifted on and off by just six men. Completely devoid of holes, the body pre-dated the similar method of making aluminium monocoque Formula 1 chassis by more than 25 years. There was no need for a radiator inlet as cooling provision was from a 75-gallon ice tank. The unusual engine layout meant the car required two three-speed gearboxes, manufactured to Railton's drawings by David Brown & Sons of Huddersfield, Yorkshire. When complete, the twin-engined *Railton Special* weighed 7,000lb, an impressively low figure considering the

John Cobb and his new wife, Elizabeth, pose with the final incarnation of his land speed car, now called the *Railton Mobil Special*, at Brooklands in 1947. *Courtesy of Charles Taylor*

Cobb explains the dashboard to his wife. At first glance the instrument panel looked massively complicated, but in fact it was duplicated side to side so that each engine could be monitored separately. *Courtesy of Charles Taylor*

CRUSADER

Bonneville 1947 — and the *Railton Mobil Special* is on the salt. So futuristic was the car that Ken Taylor heard one woman say to her friend, "Let's go around the back to take a look at the front." Nothing like it had been seen before at Bonneville.
Courtesy of Charles Taylor

car's enormous size and power. Twenty-eight companies gave services or parts to the project, Cobb himself financing the rest. It was a masterpiece of design and packaging, on par with today's Formula 1 cars, and its extraordinary body shape was like nothing seen before

Cobb arrived on the Bonneville Salt Flats in August 1938 with his *Railton Special* just after Eyston and his *Thunderbolt*. Having heard that *Thunderbolt* was failing to trip the photo-electric timing eye, Cobb had the flanks of the *Railton Special* painted black before his first run. There were only a few minor teething problems, mainly to do with running at Bonneville's relatively high altitude (over 4,200ft) and trouble with the cork floats in the carburettors, and these were tuned out with alterations to carburation and ignition. High-speed trials showed up airflow pressure dents in the tail section, so wooden reinforcements were screwed to the inside of the body.

Railton considered the potential of his car to be 350mph. On 15th September 1938 he was proved right when Cobb set the land speed record at 350.20mph.

With that, the *Railton Special* departed for home but Eyston stayed on with *Thunderbolt*, raising the record further to 357.20mph the following day.

Cobb's team returned a year later, with sponsorship from Gilmore,

Reid Railton contemplates the two Napier Lion aero engines. By constantly measuring the temperature and humidity on the salt flats, he was able to calculate the exact amount of fuel and ice to be put in the tanks, saving as much weight as possible. *Courtesy of Charles Taylor*

Railton sitting in his creation. His precision and patience made him the perfect man to take care of the warm-up procedures for the engines, obtaining just the right temperature to get the oil circulating, but not too hot so that the ice-cooling system would be brought into use. Just behind Railton, between the wheels, the camshaft-driven pulley for the anti-stall device can be seen. *Courtesy of Charles Taylor*

an American oil company. Railton had merely altered the gearing for the superchargers, but it was enough for a new record of 369.70mph, set on 23rd August 1939.

As the world returned to some form of normality after the Second World War, Cobb also decided that things should "seem like all this horrible business hadn't happened".[34] With new backing from oil giant Mobil for one more attempt on the land speed record, the car — now known as the *Railton Mobil Special* — was recovered from her safe refuge in a country barn and reunited with her engines, which had been stored at Thomson & Taylor, alongside Sir Malcolm Campbell's Rolls-Royce R-types. Reid Railton was instantly sought out and within days had a job sheet "a mile long", according to Reg Beauchamp.[35]

Pre-war, the car had suffered from stalling between gear changes, as aero engines lacked flywheels. Railton's lateral thinking soon had that problem solved by designing a clever ancillary drive to keep the engines running, manufactured and fitted by Napier's, who also completely overhauled the engines at its works in Acton, London. The axles and gearboxes, however, were found to have several hairline cracks. The new parts were made up under the supervision of Railton's friend at British Aluminium, Dr Ewan Corlett, who was an engineer from the same mould, and an expert in metallurgy. Corlett had the casings remade in a lighter but far stronger RR 53 alloy, which allowed different ratios to be installed front and rear.

When Cobb's team returned to Bonneville in August 1947, they found the

Speed hero: pictured in 1939, John Cobb looks remarkably relaxed for the camera as he poses in his land-speed car, which at this time was sponsored by the American Gilmore oil company, owner of the Mobil brand. *Courtesy of Charles Taylor*

salt in terrible condition but persevered. When the old pre-war carburettor trouble resurfaced, Railton suggested an alteration to the ignition timing, and a half-degree nose-up attitude to the body, to improve engine running and straight-line stability. He took hourly temperature readings that he recorded in his famous little black notebook and used them to make a quick-reference chart, based on estimated speed, time of run and ambient temperature, so that the minimum amount of ice was put in the cooling tank, instilling in the team, "Every half pound is ¼ of a mile an hour, and I'd like every mile an hour we can have please."[36]

On 16th September, the land speed record fell to the *Railton Mobil Special* at 394.20mph — and Cobb had broken the 400mph barrier on one of his two runs. Such was the innovation that Railton had designed into this extraordinary car that 13 years passed before its recorded top speed was beaten, by Mickey Thompson, and even then only as a single run. Cobb's achievement was honoured soon after when he was awarded the Royal Automobile Club's Segrave Trophy.

Even before preparations had started for this last of John Cobb's three land speed records, his mind had taken the same route as many land-borne record-breakers, towards the water. In an interview with Laurie Sultan, Castrol's press agent, Cobb explained why:

> Well, in 1948, moreover, after I'd broken the Land Speed Record and really achieved my ambition of obtaining a speed of over 400 miles an hour, I didn't think that I would take the car out again unless somebody broke the Land Speed Record and I took the view that they wouldn't do it for some years, and in that respect I have proved to be right for once! And after a while I thought that there was interest to be gained in high speeds on water. So I originally approached Mr Reid Railton who had designed the record car for me and asked him if he would consider the question of evolving a boat that would go very fast on water, and he told me that jet propulsion which is available of course today, is far more efficient than the propeller and that was how, and why we decided to use jet propulsion and so we approached the De Havilland Engine Company with a view to getting them to loan us an engine. The potential offer, of a loan from the De Havilland Company told us, it was a distinct possibility, and when Reid described his ideas for the craft, the die was cast then I'm afraid![37]

CHAPTER 4
REVERSING THE 'THREE-POINTER'

By the spring of 1949, Reid Railton and John Cobb were looking for a boatbuilding company to realise the speed king's new ambition to attack the water speed record and the next stages of the designer's innovative thinking, as originally conceived to solve the problems facing Sir Malcolm Campbell in 1948. They turned, logically enough, to Portsmouth-based Vosper. Having originally built Sir Malcolm Campbell's *Blue Bird K4*, the company had then been employed to convert it from piston engine to jet power and therefore had some experience of jet propulsion on water. However, this conversion had been more a question of modification than of all-out design and in reality both Campbell and Vosper's Peter Du Cane knew that the modified craft was piecemeal at best, and very unstable.

Thomson & Taylor's Reg Beauchamp, as always with a Railton project, was heavily involved from the outset:

> It was October, 1948, when Reid's head popped around the door, asking me if I had a moment. He told me that Cobb considered the land speed car was at her very limit, and although they had discussed a new car, Reid even showed me three sketches of a replacement, Cobb had decided upon the water. He had, said Railton, seen it as a sort of tradition, after Segrave and Campbell. Anyway, he asked me to give the project a code number, and to keep it as quiet as possible. By December we had the likes of De Havilland and Corlett [Ewan Corlett of British Aluminium] visiting, and by January 1949 we had a tentative

agreement with Vosper to build it. I was an engineer, and Reid told me to treat it like any other engineering problem. Once we had the solutions, we could go to Vospers.[38]

By March 1949, Beauchamp's small team had turned hours of discussion into some reasonably detailed outline drawings:

> Reid would fly in, and look at you from under the brim of his hat, that was my cue to give a quick progress report. All the time I was speaking, he'd be standing there, pulling out neatly folded pieces of yellow paper. As ideas flashed into his head, he would commit them to a yellow, A4 notebook. If these were for his use, they stayed attached, but if they were for delegated work, they would be torn off, folded, and placed in various pockets. Then, as he walked around, saying good morning, his hand would dart to a particular pocket, where he knew he had 'filed' the folded yellow page, that that particular person needed to have. It was like watching a magician at work. So, as I delivered my progress report, so, one by one, these yellow notes would appear on my desk!
>
> Reid had been in communication with [Douglas] Van Patten, while in the States, and was fascinated by his design concept for a high speed boat, so during the design meetings, Reid appeared to be talking for two people, himself, and Van Patten, but oddly, from two quite separate directions. Sometimes it appeared like he was arguing with himself.[39]

By the end of March, the tea-fuelled discussions between Railton, Cobb, Corlett and those at Thomson & Taylor had led to a concept with the necessary potential. Railton had suggested a 'reverse three-pointer' with a stepped forward shoe to aid getting the hull unstuck from the water, and to save as much weight as possible, he also envisaged remote starting batteries and refuelling after each single run.

Railton rarely, if ever, gave interviews, but an article he wrote for *Motor Boat and Yachting*, dated November 1948, goes a long way to explaining the design philosophy behind Cobb's proposed record-breaker. In fact the article was not published, probably at Railton's request because he would not have wanted either to single himself out from others in the project or to pre-empt Cobb's announcement of it when he was ready. In the article, Railton wrote:

REVERSING THE 'THREE-POINTER'

Douglas Van Patten contemplates the construction of one of the *Miss Canada* craft. Reid Railton took to Van Patten instantly, both men bouncing ideas off each other that most people would have considered outlandish. *Private collection*

To raise the water speed record any appreciable amount, and, therefore make it worthwhile, we should be aiming for over 200 M.P.H. It becomes obvious that a radical change is necessary, in both hull form and motive power. The three-point suspension, Apel-type hulls which have been used so successfully on both sides of the Atlantic in recent years, do however suffer from one major drawback, and this is a tendency to become airborne. With their curved deck surfaces, these boats bridge the gap, between a hydro-dynamical dream, and an aero-dynamical nightmare. At speeds above 150 M.P.H. the lift created on the fore-deck is such that any small change in the angle of incidence due to a wave would increase the lift so much, that the boat would take off and possibly turn over.

There are of course difficulties with propellers in high-speed craft. The efficiency of which is inversely proportional to its speed through the water, so at speeds over 200 M.P.H. the losses are so great that a very big engine would be required.[40]

Reading Railton's words, it becomes easier to appreciate the reasoning for the final shape of the proposed craft. If we follow the lines of development from Railton's original idea, certain areas of the design were down to his uncanny instinct for anything to do with speed, hence his entirely new approach to the problem. Put simply, to stop the craft taking off, it was essential to have an aerodynamically stable hull that would not be sensitive to small changes in its angle of incidence. And, quite reasonably, if the conventional 'Apel three-pointer' was so wide at the bow as to cause lift, then make the bow narrower by reversing the layout. Following on from this, a jet engine was the logical choice, because of its light weight, high-speed efficiency, and the elimination of all the problems created by having a propeller.

Some have tried to claim that Vosper proposed the reversed configuration, or that the company had been thinking along the same lines, but there is no evidence of this. In any case, why would Vosper consider such an exotic design? It hardly lent itself to high-speed motor torpedo boats — the company's main area of business.

But what had triggered Reid Railton's thinking? It goes back to 1935, when Cobb was at Bonneville attempting a 24-hour record with his *Napier-Railton*. Railton was already filling his sketchbooks and notebooks with ideas for the twin-engined land speed record car, his main aims being to eliminate wheelspin and aerodynamic lift. As has been seen, four-wheel drive cured the problem of wheelspin, but Railton felt the body shape alone was not going to ensure aerodynamic neutrality, and so he was thinking of ways to alter the car's attitude while it was running.

To this end he met with American brothers William and Larned Meacham, who in 1910 had patented a hydraulic control system to alter a hydroplane's planing surfaces while it was moving. At the time, Railton was thinking of a suspension system for Cobb's *Railton Special* based on their concept, one that would actively hold the car at the required attitude, an idea that predated by a considerable margin the active suspensions used in Formula 1 by Lotus, Williams and others in the late 1980s and early 1990s, although the technology of Railton's time was not as fast as his thinking.

Whereas conventional hydroplanes had, and still have, wide frontal areas that create lift, the Meachams wanted to counter this by having one planing surface at the front and two at the rear, on outriggers, in a form they described as a 'Canard Hydroplane'. To further counter lift at the stern created by the outriggers, they came up with a system of constant hydraulic adjustment and

REVERSING THE 'THREE-POINTER'

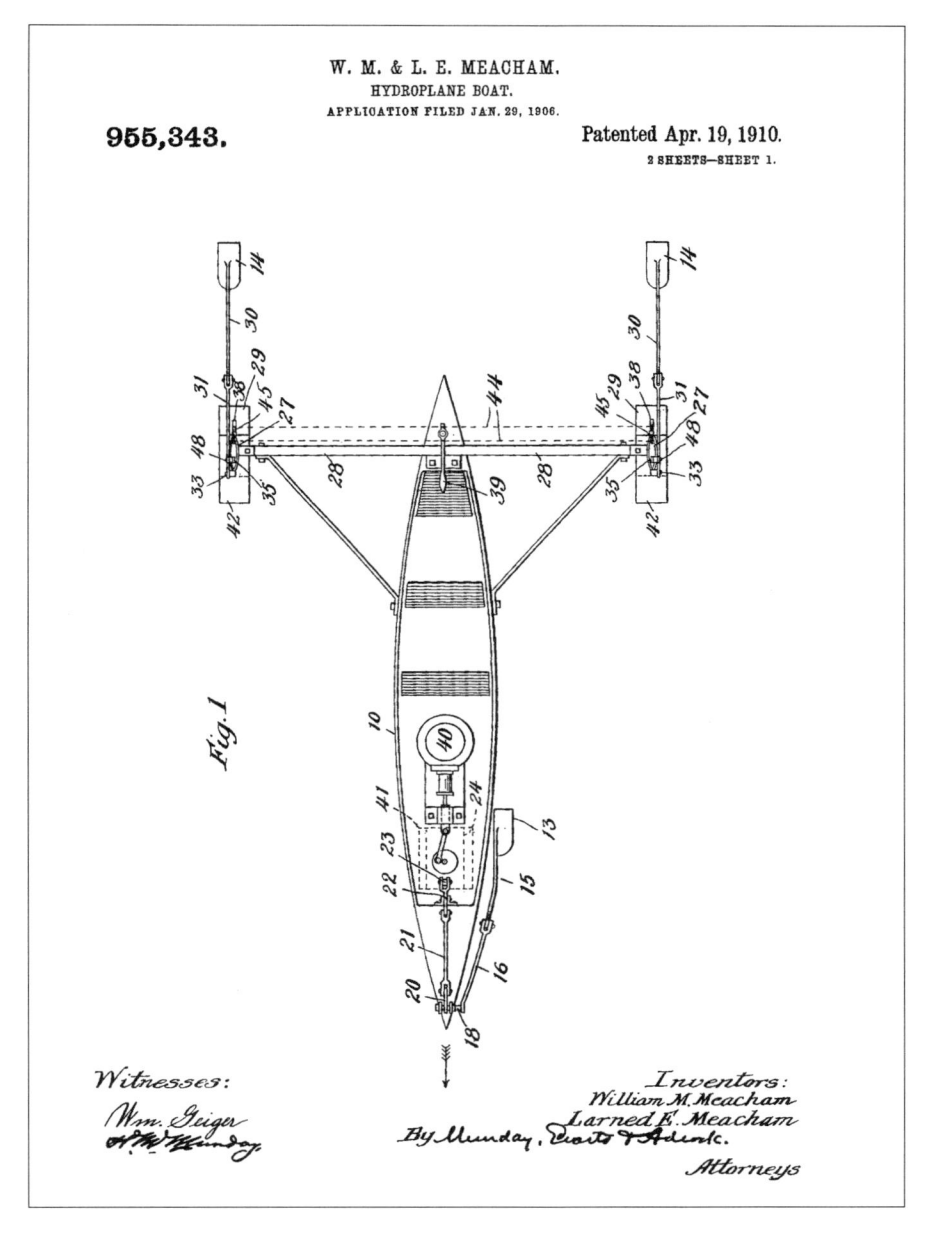

This is the 1910 patent for the reverse 'Canard' three-point hydroplane conceived by brothers William and Larned Meacham. This design first came to Reid Railton's attention because of its automatic adjustment of the rear planing surfaces, an idea that he thought he could adapt for use on John Cobb's land speed record car, but the entire concept was to play a more important role in later years. *Courtesy of U.S. Patent office/National Archive*

experimented with it for 12 years, ultimately without any great success. It is not difficult to see that when Railton needed to solve the problems of frontal lift and how to place the jet pipe of a jet engine in a water speed record boat, his earlier meeting with the Meachams pointed to the answers.

Much thought was also given to the safety of the pilot, not least as John Cobb was 'one of the family', but consideration also had to be given to the fact that they were working in post-war Britain, and materials — especially those considered 'exotic' — were difficult, if not impossible, to come by. As the structures even of jet aircraft of the period contained a considerable amount of wood, British Aluminium's Ewan Corlett suggested that Railton's 'central armature' idea, as used in *K3* and *K4*, be adopted once again, but with the use of lightweight, aircraft-style, aluminium-alloy bulkheads. Reg Beauchamp:

> I'm pretty sure Corlett had been on the telephone, not actually in the office. Reid was preparing to return to the States, and he was sitting in his meditative pose, rocked back in his seat, hands cupped behind his head, seemingly looking up at the corner of the ceiling. We were used to seeing this of course, and you didn't speak! The next stage would be sitting forward, and picking up a pencil, which either went to the paper, or to the top lip (never in the mouth!), where it would be gently tapped, until he'd start sketching. Reid was more than capable of doing the calcs but his mind fired so quickly, three or four subjects at once, so sometimes these sketches would appear on the corner of your desk, and you automatically knew you needed to stress, calculate and draw. The frame sketch looked like a skeletal tube, but Reid's preliminary outline for the hull was quite unlike anything I had seen. Two dagger-like floats at the stern, attached to a sort of flying boat hull. The cockpit right in the bow, and a stepped front shoe.[41]

One such sketch was the armature for Cobb's projected boat. Beauchamp confessed that when he saw it, he had no idea what it was. Railton knew the areas where strength was crucial, so conceived the idea of three interlinked vertical alloy hoops. The rear pair would surround and mount the engine, as well as providing the support for the rear sponsons by combining the mounting arms as part of the structures. These two hoops would be connected horizontally by alloy beams, forming a sort of tubular frame around the engine. Once Cobb had been shown the sketch of the general arrangement, it was

REVERSING THE 'THREE-POINTER'

decided to seat the pilot between the rear-mounted engine and the front shoe area, within the horizontal beams that continued to a third forward hoop. This gave a strong alloy frame around the cockpit area, protecting the pilot, as well as providing a strong mounting area for the rear of the forward shoe, the area that would, by its very nature, be the most stressed point of the vessel. As Cobb had commented that driving his land speed car — with its far-forward seating position — had been a little unnerving at times, Railton's scheme solved that problem as well, while creating others, and altering the layout considerably.

It was Beauchamp who then suggested that if the rudder was to be placed forward, as opposed to on the stern, behind the front shoe, it could be mounted on the lower, horizontal beam: "Once Reid explained the concept, it was obvious; it just didn't look very boat-like! I suggested that as the forward frame was going to be getting all the initial loading, then it would make sense to use the same mounting area for the rudder, as the drag on it was likely to be equally as high. A wry smile grew on Reid's face, 'I like that', and with that he placed his hand on my shoulder, which simply meant, well done. There was never any problem with things like that, we worked as a team, it was *our* design. He loved what he called 'the art of packaging' and if one component could do two, three or four jobs, then that was better still."[42]

For the central frame, Ewan Corlett suggested an alloy known as DTD 610B, which would give a strength of 32 tons per square inch for the structure. However, he was concerned about how the traditional boatbuilding materials of solid timbers and plywood could be successfully fixed to it.

In previous accounts[43] of the design and construction of Cobb's *Crusader*, an often-quoted letter written on 25th April 1949 by Railton to Peter Du Cane has been seen as the very start of the project. However, when the letter's entire text — as opposed to the extract that is always reproduced — is studied together with a previous letter from Du Cane that had prompted Railton's response, its contents make far more sense to the final design. The following extract from Du Cane's earlier letter, dated 13th April, quite clearly shows that Thomson & Taylor was much further into the project than Vosper. Indeed, Du Cane stated that he had only just agreed to work on the project with Cobb and Railton, and was still looking for data on more conventional hull designs, *before* Railton described what he had come up with. Du Cane wrote:

> John Cobb in the meantime is very keen still on the jet proposition and [I] have agreed to work for him with you on this problem, as Donald

[Campbell] is only really interested in the piston job.

It is a trifle difficult to know how to proceed on the jet proposition, but I think a real analysis of the forces setting up the porpoising will have to be undertaken but it is a problem requiring a good deal of time and many experiments to the model tank.

This is difficult in this country as all the tanks are so full of work.

Another line is to ascertain whether there are any forms of hull less prone to porpoising.

My first experiment in this direction was no great success, as I produced a fairly conventional type of hard chine hull and tried it in the tank as a free model with the jet unit. It porpoised much worse than the *Bluebird* and became airborne at an early stage.

Harry Greening wrote to the press here and to me on enquiry on the subject of *Miss Canada*, which is alleged to be very fast and seems to be a multi-step type with two chine lines each side.

Are these hulls really fast and steady? How do they behave? It is always difficult to get much from the press because there is so much inaccurate or partially accurate ballyhoo.

If you can get some real lowdown on these hulls, and some idea of hull form or lines this would be interesting, especially if it is a good damper of incipient porpoising.

Do let me know anything you think will be of use, and I shall look forward very much to seeing you again one of these days.

John Cobb seems very keen and I only hope I can make some sense for him eventually. One thing we are beginning to know is a hell of a lot of what not to do.

P.S. Saw Gawn [Richard Gawn, superintendent of the water test tank at the Admiralty Experiment Works at Haslar] to-day, he is going to submit to the Director of Naval Construction that a thorough research into porpoising be undertaken.

I think it will go through.[44]

While Vosper was still struggling with the problems that arose from the converted *Blue Bird K4*, Railton's reply of 25th April shows that his design work was well advanced and that he was all but fixed on the layout *before* sharing it with Du Cane.

REVERSING THE 'THREE-POINTER'

I was very glad to hear from you again and to learn of your interesting experiences with the *Bluebird* tank tests. I must say that you and Gawn seem to have done a most comprehensive job on the thing.

I also have heard once or twice from Donald Campbell, and had guessed that you might be finding yourself in an embarrassing position. In his last letter he asked me what I should advise him to do. As this was a little embarrassing also, in view of my association with John Cobb, I told him to read through the files of my correspondence with his Father, where he would find my ideas pretty clearly stated.

Judging by John Cobb's recent letters, I think there is no doubt he is really keen about a jet-propelled boat, and when once he makes a decision there is no-one more tenacious and determined to see a thing through to the end. I do hope therefore that the three of us will be able to get together on some line of action which we all agree on to be a sound one, and which will at the same time make enough sense to both the Admiralty people and also to de Havilland's (or some other engine firm) for them to give us the necessary support.

The whole point, of course, is what this line of action should be. So long as Malcolm was alive and in possession of a hull in which he had confidence, it was perhaps reasonable to persevere with it so long as he was prepared to pay the bill. As things are today, however, and having so to speak a clean slate to work on, I do think it would be a waste of time and energy to do any more work on that particular hull.

I was most interested in what you had to say about both the *Bluebird* and the hard chine models becoming airborne under test. It seems to me that this is the whole crux of the matter. I think it would be foolish for us to embark today on any project aimed at less than 200mph, which would mean designing for aerodynamic safety up to at least 250mph. If you agree on this, it seems inevitable that our first line of approach must be to think up some 'body-shape' which would combine the necessary aerodynamic properties with adequate accommodation for the man and the machinery, and then (and only then) to try and work in the necessary buoyancy, planing surfaces, etc.

The sort of thing I have in mind as I write is a very small sea-plane hull with a couple of small floats or 'skis' mounted on outriggers where the tail normally is: i.e. a tricycle with the one wheel in front. Probably it might be nothing like this, and I mention it only to show

how radical I believe we shall have to be before we are doing 200mph, with confidence [*author's emphasis*].

I hope all this will not sound too much like nonsense to your expert ear. I know how easy it is to sit in a chair and make suggestions. I have suffered from that myself! I realise that anything we do must appear 'reasonable' to Gawn, or we could not get his quite indispensable help. Furthermore, if he wouldn't help, we could hardly hope for cooperation from the engine manufacturers. On the other hand, the more I think of it, the more reluctant I feel to advise John Cobb to embark on any project where the aerodynamic limitations are so severely critical as they must be with any conventional type of hull that I have ever seen or heard of.

I think I have said enough to show you the stage that I have arrived at in my own thinking — such as it is. The two things which seem to me to be of immediate importance are (1) whether your Company would care to be associated with a venture which might lead us away from conventional boats and into something more like the aeroplane field, and (2) whether the Admiralty Experimental Station [*sic*] would consider such a venture to be of sufficient interest to warrant their continued support. A third point would be, of course, the cooperation of an engine manufacturer, but I feel sure that with the combined prestige of yourself, Gawn and John Cobb, we could soon pry an engine loose somewhere.

You ask me about Harry Greening and his *Miss Canada*. As it happens, I know the designer of this little boat — Douglas Van Patten — quite well, though I haven't seen him since he and Harry G. and I spent an evening together last year. He lives near Detroit, and I will look him up and ask him a few leading questions when I go back there.[45]

In reply on 4th May, Du Cane clearly showed that Vosper had not been thinking along anything other than conventional terms, and he had taken Railton's reference to 'skis' quite literally:

Thank you for your interesting letter of April 25th. I agree with what you say and will investigate closely the possibility of designing something of true aerodynamic form running on some form of skis.

REVERSING THE 'THREE-POINTER'

Miss Canada IV at speed. The hull incorporated much of Douglas Van Patten's revolutionary thinking, especially below the water line, and it was this craft that brought Railton and Van Patten together. *Private collection*

> Whether this will be permitted is another matter, but I certainly agree that at first sight it seems to be a good line of approach.
>
> John Cobb is extremely keen and I am glad to say that Gawn is still prepared to work for us, having been given permission by the Director of Naval Construction.[46]

Back with Railton's team, Van Patten's revolutionary influence was clear to see in the preliminary drawings, in which Railton envisaged a seaplane type of hull to give the necessary buoyancy, with the rear sponsons outrigged, producing what he described as "the tricycle effect", with one 'wheel' at the front, and two at the rear.

Indeed, Railton did not hand over the initial Railton/Van Patten model and blueprints to Vosper until March 1950, as confirmed by a letter he received from Cobb acknowledging receipt of his copies of the drawings that Cobb said he would "take down to Du Cane after he returns from his holiday".[47] The letter is dated 8th April 1950, just before Railton's letter to Du Cane giving him the broad outline of what to expect, although the 'forward cockpit' model was kept back.

CRUSADER

In a note written to John Cobb at about the same time, Railton also stated:

> I have a verbal agreement with De Havilland's for a late Goblin or a Ghost, so I now have an idea of dimensions. These radial flow jets are a bit broad in the beam, but [Frank] Halford [technical director of De Havilland's engine division] promises it will be FOC [free of charge] through the Ministry. As for Vosper, it's really our only port of call for the build, so to speak. Saunders-Roe, although constructing flying boats, and dabbling with jets, have no experience of a jet in a motor boat. I had a good long chat with Peter Crewe [chief hydrodynamicist at Saunders-Roe] and he tells me he is happy to give us whatever help we need from his flying boat experience. Vosper are unique in that with Malcolm's boat, they have built a jet hydroplane, I just wonder how they will take to our ideas. Even their high-speed craft are traditional designs and quite conservative when it comes to design, materials and construction. The Van Patten model will suffice for the time being as I believe we really do have something quite different from the norm.[48]

As if to reinforce the need to be revolutionary in design to reach the project's goal, an earlier letter from Railton to Du Cane, dated 19th November 1949, clearly stated that 'conservatism' would hold the project back, and therefore the speeds attained:

> This is to inform you that I have just mailed back to you the Report on the meeting of the Tank Superintendents [at the Admiralty Experiment Works at Haslar] which you kindly lent me and which I have read with much interest — particularly in so far as it reveals the aggressive conservatism of our friends at Haslar.
>
> I have been thinking a lot about our proposition, but I am well aware that no further observations on my part can be of the slightest use at the present stage, and for this reason I have purposely refrained from troubling you with any.
>
> I have been seeing a good deal of Douglas Van Patten and his boat *Astraea II* — a sister ship of *Miss Canada*. The more I see of this 3-step design the more I am convinced of its particular adaptability to our proposed shape of vessel. I am not suggesting that this conviction of mine should influence your present line of investigation, which may

REVERSING THE 'THREE-POINTER'

Reid Railton sits in another of his creations — Goldie Gardner's MG EX135. The fully enveloping bodywork proved to be extremely efficient, even though Railton and Reg Beauchamp based the shape solely on their slide-rule calculations, without access to a wind tunnel. *National Motor Museum*

well lead to something better still. However, I do feel that, if we encounter unforeseen difficulties on our present lines, we do have this to fall back upon as a well-tried form which, by pure chance, happens to fit our desired aerodynamic shape almost perfectly.

There seems to be no doubt that *Miss Canada*, with an alleged 2,250 hp, has twice exceeded 150 mph (electrically timed). On both occasions the bronze propeller bent under the load, so that the boat was unable to make the return trip.

I have got some excellent stereo-photographs of the under-water form of *Astraea* which I look forward to showing you some day.

Don't bother to reply to this — I know I shall be hearing from you when you have anything to report. I expect to be here until the middle of December, and thereafter at Berkeley [Railton's home in California].[49]

It is clear from Railton's communications that he was trying to steer those outside the Thomson & Taylor part of the project towards what he and his team had established, albeit only in sketch and model form, while also being careful to be gentle with the egos of people at Vosper.

It should be said that no legal or official contract appears to exist in the archives of any of the parties involved. Although all were gentlemen of the era, it is most unlikely that the agreement would have been made on just a handshake, but the lack of a copy of any contract is puzzling. Indeed, in early correspondence Du Cane merely confirmed Vosper's "support", with Cobb to cover all costs, both of construction and running. If this was indeed the case, then it becomes obvious that, without defined roles and responsibilities, there was ample opportunity for some to overstep their marks. Having said that, all of the correspondence from Railton and Cobb to Vosper is very much in the vein of the 'client' speaking, and that from Vosper of the company employed to carry out the client's instructions.

Undoubtedly Du Cane considered himself the 'designer' and Railton was constantly referred to as a 'consultant', but, as we shall see, this irked both Railton and Cobb. This lack of clarity helped lead to some breakdowns in communication regarding the aims of the project as well as allowing some to see their own positions differently from what was intended.

Reid Railton had a lot on his plate at this time. He was travelling frequently between the United States and Britain, and overseeing several projects on both sides of the Atlantic, although his friend John Cobb's new contender for the water speed record was centre stage. However, as various problems arose with Cobb's project, Railton was having to spread himself very thinly indeed. During early development of Cobb's jet-powered craft, the pressures on Railton were keenly observed by Zillwood 'Sinbad' Milledge, a notable Brooklands engine tuner who spent a lot of time with him. Milledge stated:

> I remember the first time I met Reid, possibly at the Hermitage, and we got on well, and he wasn't shy about sharing information, which endeared him to me, racing could be a little secretive shall we say. We had a rapport, engineer to engineer; we told each other things, knowing the other would not repeat it, it made a very pleasant change. Because of this, it became second nature from then on, to pop in, say hello, chew the fat, and get on with the day.
>
> I recall going into his office, over in the [Brooklands] paddock, and someone said, 'Not now, headache'. I had no idea what this meant, but

REVERSING THE 'THREE-POINTER'

Reid Railton's work for Hudson — brought about by use of a Hudson Terraplane chassis in the Railton road car — was one of the reasons why he moved his family to the USA. This rather staged photo has Railton extolling the virtues of a Hudson model to John Cobb on the Bonneville Salt Flats in 1947. *Author's collection*

I was later told he could get these nasty headaches, very short notice, and I walked in to see Reid banging his forehead with the base of his palm, really quite hard. I left without saying anything, but later Reid said it helped clear them, so no more was said.

Anyway, years later, and he was back and fore between England and America, as well as working with Freddie Dixon and Raymond Mays on a project, assisting Donald Campbell amongst so many other things, and he looked tired, and if it wasn't a headache, you'd often see him pushing his wrist into his eyebrow while squeezing his eyes tight closed. Three minutes later and you were none the wiser. I don't know how he did it.

He never carried a briefcase, but there was THE black notebook (everyone knew about THE notebook!), and every pocket had folded pages of yellow paper, all seemingly put in certain pockets in some sort of system. He knew what was on each sheet, and which pocket it was in and who it was for.[50]

CRUSADER

As an indication of just how busy Railton was at the time, he was also advising Major Goldie Gardner on his continuing record-breaking exploits with his highly successful and long-serving MG EX135, for which Railton had designed streamlined bodywork back in 1937. Indeed, at one time every speed record — for engine size or distance — was held by either Cobb or Gardner in cars fully or partly designed by Reid Railton. In America, where Railton now lived, his involvement with Hudson on the Railton car had led him to accept a design consultancy role with the Detroit-based manufacturer, in order to help develop the handling of its Terraplane model to European standards while also refining the transmission and drivetrain.

Added to Railton's workload, while he was in the United States nearly all his communications with Cobb, with his colleagues at Thomson & Taylor and with Vosper had to be done by airmail or telegram, as telephone calls at that time were generally limited in duration and often had to be pre-booked up to three days in advance. Relying on airmail meant that a simple 'conversation' on a matter could become spread over as long as ten days — but a helpful consequence is that we have been left with a wealth of archived correspondence.

CHAPTER 5
TENSIONS RISE

Vosper, and in particular Peter Du Cane, held great stock in testing with scale models. In one note to John Cobb, Reid Railton stated: "I'm sure we can glean a lot of data from these model tests of Vosper, but the impossibility of scaling the medium (water) is going to cause us a deal of problems later on. We need to move through this stage with some haste. There have been some good results with the 'egg on skis' [an early Vosper model], but we already knew the positions of the planing surfaces, and the layout from our work with Van Patten, and the Apel formula, but one assumes Peter needed to see it for himself rather than take our word."[51]

One other drawback with relying on these scale-model tests was aerodynamic and hydrodynamic cross-over. Vosper was producing various models of differing scales to test different environments in different ways. Whereas today we have wind tunnels with rolling roads and even moving water that can test simultaneously for both 'media' interaction *and* aerodynamic influence, Vosper was attempting to apply wind-tunnel data of one scale to water-tank data of a different scale, with one 'medium' — water — that cannot be scaled. This also applied to the free-running, remote-control models with approximations of the likely jet thrust lines and available power that were run on open water, sometimes in less-than-perfect conditions. When dealing with new and untried technology, it can be argued that these were the best methods available, but as Railton commented to Cobb: "It's a good base, but there seems to be an over-reliance on these scale tests, which simply don't seem to match up, and to my mind, seem a little rushed into the bargain."[52]

CRUSADER

Regardless of this, Railton and his Thomson & Taylor allies continued to develop their concept (based on the calculations done thus far) and their own model (based on Railton's original sketches and discussions with Douglas Van Patten).

Meanwhile, Sir Charles S. Lillicrap, Director of Naval Construction, gave permission for use of the huge towing tank at the Admiralty Experiment Works at Haslar, near Gosport in Hampshire. Here Vosper started testing for the Cobb project with $1/16$-scale, rocket-propelled models of around 2ft in length. Initially a rough cigar shape was chosen because alterations made in the angle of incidence would move the centre of lift to behind the centre of gravity, providing a self-righting dynamic, and an air rudder was used, solely to keep the model straight during the tests.

The first major problem to overcome was that of porpoising. With the jet-engined *Blue Bird K4* in which Sir Malcolm Campbell had made his abortive attempt on the record in 1947, the porpoising took place at speeds of less than 100mph, proving that the cause was hydrodynamic rather than aerodynamic, as Du Cane and Vosper thought at the time.

As a consequence, Vosper's first model had four skis disposed in a diamond shape. There was one on either side to provide lateral stability, while the forward and aft skis were intended to stop any tendency to porpoise. The purpose of the skis was really only to replicate the planing surfaces. It soon became apparent that the rear ski was unnecessary, and so the 'egg on legs' model, resembling a water tricycle, came into being.

A considerable amount of data was collected about the positions and angles of the skis, none of which contradicted the calculations done by Railton's team at Thomson & Taylor. However, the model, by its very nature, had to be launched from a ramp, and so a core problem remained — how to lift the real boat out of the water from rest and up onto the planing surfaces. It was obvious at this stage that the skis needed to become floats, with the forward surface faired into the hull, very much on the lines of Railton's first sketches.

Tests continued with a $1/6$-scale model to get an idea of its hydrodynamic/aerodynamic interaction. This was powered by a rocket motor of 30lb thrust and a burn time of 20 seconds, made and calibrated by the Rocket Propulsion Division of the Royal Aircraft Establishment at Farnborough. The actual speed of this early Vosper model, 97mph, equated to a scale speed of 240mph.

It is unclear exactly what tests Cobb's team was conducting at this time. They were either unable or unwilling to unveil their own version, probably

TENSIONS RISE

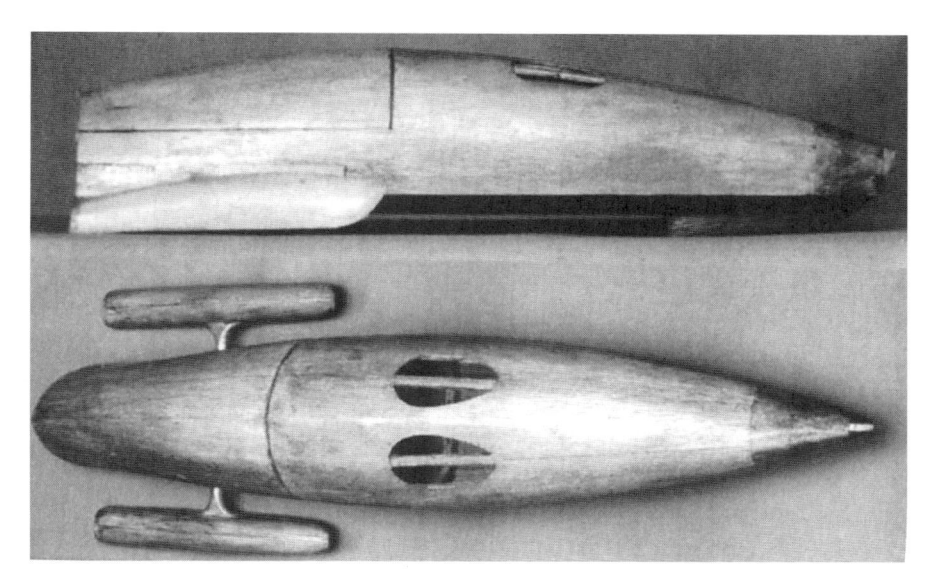

An early concept, showing Vosper test model CJK in June 1950, probably in its 'E' variant with adjustable rear sponsons and the very narrow, short forward planing wedge that refused to rise onto the plane. The last version of this model had the complete Railton/Van Patten two-step forward wedge, with a small degree of vee and raised rear sponsons. It resembled the final shape much more closely and got 'over the hump' without problem. *Courtesy of BAE Systems (formerly Vosper Thornycroft)*

because Railton wanted substantial, accurate data to counter any perceived errors in data relating to Vosper's model. Either both teams' data would agree, or at the very least they could amalgamate the best of both designs.

Vosper persisted with the testing of its models, even though Railton predicted certain problems. Although at first these were dismissed by Du Cane, they proved to be correct, as the Vosper man had to confess in a letter to Railton dated 3rd January 1950:

> I think you actually owe me a letter, but it is probably on its way.
>
> However, I have had a certain amount of running of the models since I wrote to you and have also taken the bi-fuel rocket motor project a stage further.
>
> As far as the models are concerned, I cannot say that I have had any very good running as yet, as the latest model unfortunately does not un-stick, which is precisely what you predicted, and what I erroneously

thought would not be a trouble. I have made various modifications to the nose to try to get it to un-stick, so far without success.

I do not regard this as being impossible, but it certainly is not as easy as appeared at first sight, so that I do think it wise to have the alternative of Van Patten's model up our sleeve.[53]

What is clear is that Railton, while still in America, had continued his work with Douglas Van Patten. He had an '80% Van Patten' model sent over to Vosper for testing in April 1950, as a 'second string'. Meanwhile, he had with him the Thomson & Taylor model that he and Van Patten were testing in private alongside a '100% Van Patten' design. The Railton model's forward step can best be described as a 'one-and-a-half step', with the area closest to the hull being wider and flatter than the actual planing area of the second part of the step. At a stroke, this solved the 'un-sticking' problem, as confirmed in a letter sent by Railton to John Cobb in early December 1949:

> I trust all is well with you and yours? I have been spending a deal of time with Douglas [Van Patten], and our hybrid Van Patten/T&T model. I believe a mixture of the stepped forward shoe, and its angle of attack, and the rounded cross section of the main hull, is leading to the model getting over the hump [i.e., 'unsticking'] at quite low speeds, which of course is all grist to the mill. We have no real means to calculate the scale speed from the towing rig here, but the visual results are very encouraging. Indeed, the pure-bred Van Patten model is showing much higher resistance at speed than our own. But Michigan's equipment is at its very limit.
>
> I'm puzzled then, as to why the tests at Haslar seem to be the opposite, as regards getting on to the plane, as from what I have seen, the planing surfaces seem overly large in scale, but then again, the Vosper models seem to have a vast flat area underneath. Still, between us, we should be able to combine the best data from both into a fine craft.[54]

Thus far, it has been possible to follow these written communications between Vosper on the one hand and Railton and Cobb on the other, but they do not tell the full story. With study of the archives of both Reid Railton and John Cobb, it becomes apparent that these 'discussions' formed a triangle, with

TENSIONS RISE

Vosper at the top, dealing with both Railton and Cobb, and with Railton and Cobb at the base, with a separate line of communication between them.

For his part, Cobb was more than happy to leave things in the hands of Railton, his old friend, while Railton was happy to listen to what Cobb had to say and suggest accordingly with the Thomson & Taylor part of the project, while also working with Douglas Van Patten. As Railton stated to Cobb: "Belt and braces my friend. We have the central scheme, and then we can select the best of all the tests, to tackle any problems, unforeseen or those created for us!"[55]

That last comment appears puzzling and needs to be considered in context. To do this, the best approach is to go back to the early conception, model tests and construction from the perspective of those at the coal face — Reid Railton, John Cobb and Peter Du Cane — starting with a letter from Cobb to Railton in July 1948:

> Further to our conversation the other day, I feel it only right to put in writing what has been going through my mind.
>
> I feel there is little to be gained, fine tuning the car, and returning, at no little expense to Bonneville, for an improvement of 4 or 5 miles per hour. Added to which is the apathy from Dunlops! Even if they were to play ball, I agree with you that there are few choices as to what to power a new car with, that is why I suggested that we might look toward the water.
>
> At least here, where, as you quite rightly pointed out, pure thrust overcomes the underwater problems, we might pry an engine loose. Plus, Malcolm has already set a precedent with *Blue Bird*.[56]

There is no record of a response, but as this letter was sent to Thomson & Taylor, it is likely that Railton simply telephoned or waited until the two men met up again. In September 1948, Railton wrote to Cobb:

> I have had time to sit in the dark and squeeze my head so to speak, and feel we are on the right track. Having seen the Apel *Bluebird* model take off in the wind tunnel, it's obvious all that air around the front is our enemy. It strikes me therefore, that if we turn the layout around, with the single shoe at the front, we can present a much more narrow

and more aerodynamic front end, allowing us to overcome the problem at a single stroke. This puts the remaining shoes to the stern, and by raking the supports for these, we can achieve the perfect 'wheelbase' according to Apel's formula. All we have to do then (as if it's no mean feat!), is place the engine and its intakes, and the pilot in the correct places within the layout.[57]

There is no further written record until January 1949, shortly after the death of Sir Malcolm Campbell, when Railton wrote:

Not totally unexpected news about Malcolm, he certainly didn't look well when I last saw him. I suspect a few people's lives might have become a little easier; he had become very difficult to deal with.

It seems we are pretty much set to go. The engine position is far from settled, but Frank [Halford, at de Havilland] insists it's unlikely that we won't get at least a Goblin, but more likely a Ghost with some form of re-heat?

Ewan [Corlett] tells me he will contact you directly, and does not foresee BA [British Aluminium] not coming on board.

I know George [Eyston], was keen on Thornycroft, but Vosper did at least put the Goblin in *Bluebird*, and have experience with Apel's Ventnor.[58]

The next letter is dated 14th March 1949, from Cobb. It mainly concerns a proposed visit from Railton and his children, but there is a hand-written 'PS' that states: "Du Cane thinks *Bluebird* might be bought for £1,000, which looks reasonable if an engine can be obtained. Will let you know more later on."[59]

This is interesting: Cobb, who had absolutely no experience of high-speed boats, was obviously looking to acquire some by purchasing *Blue Bird K4* from Sir Malcolm Campbell's estate. When it became clear that in fact Donald Campbell wanted to buy his father's last car and boat, Cobb looked briefly at trying to buy *Blue Bird K3* from Sammy Simpson, a Lincoln car dealer based in London, but felt the price for both the hull and the engines (which Donald would want anyway) was too high. Railton replied by saying that he did not feel either hull would give Cobb the right 'feel' as his proposed design was "considerably different from conventional craft". He added, perhaps as a joke,

TENSIONS RISE

"Maybe we can produce a smaller, prop-driven 'trainer'?"[60]

Cobb responded on 25th March 1949 with the following:

> I am very keen on the boat project so long as I have your support, but I am rather fed up with Du Cane, as I can get nothing out of him, he always seems to be away, and although I have offered to go down to Portsmouth, I have so far failed to see him.
>
> He did ring up once to say that he thought *Bluebird* might be brought for £1,000, but that he has not yet done his tank tests. It doesn't sound dear at this price if we could get the engine back. Would appreciate your ideas. I presume any design you had in mind you would want to try in the tank, and I suppose Du Cane is useful for this. I will make another attempt to see him next week.
>
> I was surprised to see that Malcolm only left £175,000. I thought it would be more. They say he died from a stroke, brought on by the sight of a bill from his doctor![61]

When Cobb wrote that letter, it seems that he may not have seen Railton's previous reply, owing to the time required for letters to travel across the Atlantic. Railton's next letter stated:

> As I previously said John, there is little to be gained, other than a dent in your wallet, in pursuing either *Bluebird* hull. The Saunders hull was very unstable, and the Vosper one has no engine, and Frank [Halford] tells me there is little chance of it [the engine] being made available again, as it flat out did not work, and it never was going too either! He also tells me that a Goblin or Ghost won't be a problem. We just have to keep it under our hat, and let others make the approaches. The Ministry needs to deal with companies, rather than individuals, after Malcolm rather dragged them through the mud.[62]

Very soon after, on 1st April 1949, Railton wrote again:

> Your laments about Peter Du Cane read almost word for word like Malcolm's used to! Young Donald wrote me the other day (in strictest confidence, which I am now violating, so don't give me away), that he had ideas of running the *Blue Bird* himself. This could easily account

for Du Cane's elusiveness. He may scent a sucker and some quick money for his company.[63]

Cobb replied on 7th April 1949:

> I can assure you I will not divulge what you say about Donald Campbell, but he has been telling all and sundry that he proposes using the boat if Du Cane's tank tests are satisfactory, but from what I see in him I should imagine that it is very unlikely that Halford would let him have an engine. He is also talking of the possibility of using the boat again with the piston engine. I was under the impression that Du Cane had said that he was going to conduct our tank tests privately and was surprised when I heard that he had told Donald. As you say I think he has his eye on the main chance, hence his elusiveness.
>
> I phoned him last week and he told me he would be completing the tank tests by the 1st April, but I have heard nothing from him since, although I have telephoned him and have been told he is out. His co-operation with Donald is bringing about the same sort of impasse as we were in last year before Malcolm died and it is all very irritating, but I am keeping my impatience under control as I realise he is useful with his contact with the Admiralty, and also I suppose he has got a certain amount of 'know how' with regard to building a hull, but I believe if necessary we could get the use of the tanks and also there must be other people who could build a hull? The main thing is I definitely want you to design a hull if you will do it, and as always I want you to have a free hand to do what you think best, and if you are agreeable I think perhaps it would be better if you did not assist Donald in the unlikely event of him doing the thing, as I think it might lead to a lot of complication. How about Thornycroft's with or without George [Eyston] intervening?[64]

From this, it appears that an agreement with Vosper to build Cobb's jet hydroplane was not set in stone, which might confirm the absence of a formal contract, and that Peter Du Cane's lack of progress was causing frustration for both Cobb and Railton. In his defence, his prime focus was to make money for his company. Lorne Campbell, a marine architect of considerable note, began to learn his trade as an apprentice at Vosper, under Hughie Jones, who was heavily involved in the *Crusader* project, and at a time when the company was still reliant on Du Cane, and his observation was as follows: "Peter Du Cane was always working hard to keep Vosper going, I think it was quite a struggle in those days. As such, he was juggling a lot of balls I believe, but it is constantly apparent, his mind was not fully on the Cobb/Railton project."[65]

Railton said as much in his next reply to Cobb, dated 10th April 1949:

> As regards Donald, I can see no conflict of interests at all. He is almost resigned to the fact that he will have to revert to the old piston engine, and that any speed he sets, will only take the record by the slimmest of margins. I have told him that he will find all he needs to know by going through the file between his father and I, and any further advice I can give is limited, as the design is now relatively 'old'. I shall of course defer to whatever you feel is the right thing to do, though the lad is struggling.
>
> As you know, we have a pretty good outline to go on from, and a verbal agreement from Frank [Halford] for an engine. Ewan [Corlett] has been fighting successfully in our corner as well. Saunders have ruled themselves out, though Peter [Crewe, Saunders-Roe's chief hydrodynamicist] is happy to advise. Thornycroft's of course built *Miss England III*, but that was a single step. I therefore feel, possibly against better judgment that Vosper, with their limited experience, albeit, the only experience anyone has of a jet in a hull, are our only option. I am more than happy as you know to get on with the pencil work, but I can sense we will become quite frustrated, and it may be difficult to keep Du Cane on the line, he has his own ideas, but I'm proposing something quite different, which may not sit well with him or his company.
>
> We do however have the option of De Havilland, they know their engine, and aircraft construction, and that is part of the route we need to follow.[66]

Meanwhile, Du Cane had written to both Cobb and Railton, drawing the following comment from Cobb in a letter to Railton dated 22nd April 1949:

> Du Cane has sent me a copy of his letter to you dated 13th April and you will observe that he is now ready to co-operate, I suppose this is the

> best plan for us now especially if you are giving the job some thought. George [Eyston] is home but I shall not say anything to him unless you think it desirable. I have a date with Halford next Wednesday, I feel sure he is very interested and I will let you know what he has to say. I shall be surprised if Donald goes through with the other proposition when the bills start to come in, Malcolm left £175,000. Death duties £77,000, £400 pa to Dolly which would absorb £10,000. Nil to the other wives!! £1,000 and a house to Villa, the remainder in trust for the two children, so he can't get more than about £1,000 pa out of it, however it's not my business![67]

In a fast turnaround, Railton replied on 25th April 1949. Interestingly, the letter contains comment that contradicts a previous reference to Vosper having come up with the idea of using small rocket motors to replicate the thrust of a jet engine in the scale-model tests at Haslar.

> I have received a long letter from Peter Du Cane telling me the results of the latest tank tests of the *Bluebird*. They did a lot of work with actual miniature jets, as I had suggested, but the results were entirely negative, i.e. they couldn't reproduce the porpoising which Malcolm complained of.
>
> By giving the rocket an extra charge, they made the model go so fast that it 'took off' and turned over at a scale speed of about 170 mph.
>
> He also told me in confidence of his recent contacts with Donald Campbell. The upshot is apparently that he is free and willing to work with us on a jet boat. However, he says quite frankly that he feels rather stumped at the moment as to how to go about it.
>
> As you know, I have fairly definite ideas of how I should tackle the problem...
>
> The first thing to find out, is whether Du Cane and [Richard] Gawn [the man in charge of the Haslar towing tank] are prepared to undertake the really radical investigation required, either on their own and with my co-operation, or under my direction, and (of course) with their co-operation. In view of my domicile here [Berkeley, California], the former arrangement would be much the best, provided they can be put in the right frame of mind.
>
> I have written to Du Cane as per the copy enclosed, in the hope that

> 2808 Oak Knoll Terrace
> Berkeley 5. Calif.
> April 25th. 1949
>
> Dear John,
>
> I have received a long letter from Peter DuCane telling me the results of the latest tank tests of the Bluebird. They did a lot of work with actual miniature jets, as I had suggested, but the results were entirely negative, i.e. they couldnt reproduce the porpoising which Malcolm complained of.
>
> By giving the rocket an extra charge, they made the model go so fast that it "took-off" and turned over at a scale speed of about 170 mph.
>
> He also told me in confidence of his recent contacts with Donald Campbell. The upshot is apparently that he is free and willing to work with us on a jet boat. However, he says quite frankly that he feels rather stumped at the moment as to how to set about it.
>
> As you know, I have fairly definite ideas about how I should tackle the problem, although I have very little idea of the naxture of the solution to which it would lead. The first thing to find out is whether DuCane and Gawn are prepared to undertake the really radical investigation required, either on their own and with my cooperation, or under my direction and (of course) with their cooperation. In view of my domicile here, the former arrangement would be much the best, provided that they can be put into the right frame of mind.
>
> I have written DuCane as per copy enclosed, in the hope that it may have some stimulating effect. For the rest, I cant do much until I arrive in August, by which time I hope to have the thing thought out sufficiently to come up with concrete suggestions if required.
>
> It might be a good thing if you can make an opportunity of going down to Haslar with DuCane to see Gawn. He is a nice little man, and such an acknowlegement of his importance to you might do some good.

It is clear that as early as April 1949, Reid Railton had definite views on what should be built, and that Vosper was "at a loss" as to what route to take. Many of the letters Railton sent show he was concerned that his proposals for *Crusader* were far from what contemporary boat designers and builders had ever considered, and feared the concept would not be accepted. *Railton/Cobb archives*

it may have a stimulating effect. For the rest, I can't do much until I arrive in August, by which time I hope to have the thing thought out sufficiently to come up with concrete suggestions if required.

It might be a good thing if you can make an opportunity of going down to Haslar with Du Cane to see Gawn. He is a nice little man, and as such an acknowledgement of his importance to you might do some good.[68]

CRUSADER

ANNING, CHADWICK & KIVER, LTD.
H. W. CHADWICK E. R. CHADWICK K. G. HOLLEBONE J. R. COBB W. P. JAMES A. J. ASHFORD
INCORPORATING ANNING & COBB, CHADWICK & HOLLEBONE AND HENRY KIVER & CO.

TELEGRAMS & CABLES:
COBANG, LONDON
HIDEBROKA, LONDON
KIVALFUR, LONDON

TELEPHONES:
MANSION HOUSE 5544 (5 LINES)
         "   4801 (3 LINES)

CODES:
BENTLEY, A.B.C. 6TH EDITION
TANNERS COUNCIL, ACME
AND ALL WELL-KNOWN CODES

AIR MAIL.

JRC/MC.

15, Arthur Street,
London, E.C.4

AND AT LIVERPOOL, NEW YORK, WINDHOEK (S.W.A.)

31st May, 1949.

Mr. Reid A. Railton,
Whittier Hotel,
Burns Drive,
Detroit, Michigan,
U.S.A.

Dear Reid,

    I have not written to you since I received your letter of the 25th April enclosing copy of yours to Du Cane. I gave him a week or two to digest it and then tried to contact him, with the usual result. However, I have got him where I want him at the moment as he is in a London Nursing Home, having had a minor operation, so I have been able to talk to him.

    My last visit was more satisfactory, as he more or less promised to get a move on, but I am of the opinion that he is going to spend a good deal of his time this summer playing about with the "Bluebird" with Donald Campbell, as he is going to put the piston engine back and have a go at the record. I feel in my bones that we are going to build this boat, but somehow doubt if Vospers will build it, but I am not going to do anything to prejudice the situation, and await your arrival in August.

    I saw George the other day and he asked what was doing, so I told him I was playing with the idea of ~~the~~ boat but have not made any plans. He said that Thornycrofts would certainly be interested and that their chief Naval Architect was well in at Haslar. On receipt of your letter of the 21st May, I wrote to Arthur Bray and asked him for the relevant documents, but have not heard from him yet. I will Airmail them to you as soon as possible.

    I am going into the question of a car for you with Godfrey Davis. Normally they only supply 10 h.p. Morris well governed down and they will only do about 50 m.p.h. which makes it very irksome for any distance. However, I will see if they can do any better. Let me know if you think the Morris

p.t.o. ⟶

In this letter from his company address, John Cobb shows that he is already having problems in discussions with Vosper, even though Thornycroft is still being considered as an option, through George Eyston, who has been brought into the project's confidence. The letterhead gives us a clue to the complicated set-up of the companies that the Cobb family owned and controlled. *Cobb/Railton archives*

TENSIONS RISE

This letter suggests that Vosper had not as yet come up with a concept of its own as to what to build, which rather contradicts published history.

Cobb's next letter was more about other matters, but his 'PS' was of interest: "I watched a young lad on his tricycle yesterday, big wheel on the front, and two little ones to the rear, and while it convinced me of the idea, I am relieved I won't be peddling!"[69]

When Railton eventually wrote again, on 21st May 1949, his only reference to the project was to express his concern that his idea for the boat was so unconventional that it could contravene the regulations. He asked Cobb, therefore, to contact Lieutenant Commander Arthur Bray, one of the British representatives on the worldwide governing body, to obtain a copy of the very latest regulations for an attempt on the unlimited water speed record.

From other communications at this time, it is also apparent that George Eyston was already involved in Cobb's project, which might indicate that his sponsor, Castrol, was interested in supporting the venture. While Cobb and Eyston shared a distaste for the limelight, they were otherwise very different characters. Whereas George could be impetuous and keen to get things done, John would never be rushed, preferring to reach his goal calmly, step by step. Regardless of these different approaches, they remained firm friends throughout their duel for the land speed record, to the point where it seemed obvious that Eyston should become the manager of the *Crusader* project.

Cobb not only had problems contacting Bray but remained unable to speak to Du Cane, as is clear from his letter to Railton dated 31st May 1949:

> I have not written to you since I received your letter of the 25th April enclosing a copy of yours to Du Cane. I gave him a week or two to digest it and then tried to contact him, with the usual results. However, I have got him where I want him at the moment as he is in a London Nursing Home, having had a minor operation, so I have been able to talk to him.
>
> My last visit was more satisfactory, as he more or less promised to get a move on, but I am of the opinion that he is going to spend a good deal of his time this summer playing about with the *Bluebird* with Donald Campbell, as he is going to put the piston engine back and have a go at the record. I feel in my bones that we are going to build this boat, but somehow doubt if Vosper will build it, but I am not going to do anything to prejudice the situation, and await your arrival in August.

> I saw George [Eyston] the other day and he asked what was doing, so I told him I was playing with the idea of a boat but have not made plans. He said that Thornycrofts would certainly be interested and that their chief Naval Architect was well in at Haslar.[70]

Days later Cobb added to this with a Telex from Anning, Chadwick & Kiver: "Reid, would you allow me to show George your drawings of the boat? He needs to see just how far we have gone to speak to Thornycrofts as Du Cane's outfit seem more keen on *Bluebird* and conventional design, which may be why he is stalling."[71]

On 21st June Cobb wrote again:

> I have found Arthur Bray very tiresome regarding the regulations you require, and from what I can gather from him there do not appear to be any, and he tells me anything is allowed except moveable aerofoils. On the other hand hydrofoils are alright and an aerial rudder is not objected to.
>
> As far as I can see, any specification that was submitted to the governing body would be examined and if thought it added to the efficiency of fast craft they would make their regulations fit it.[72]

On 20th July Railton wrote to Cobb at his Arthur Street office:

> I believe we are in good shape. Reg [Beauchamp] has produced some good figures based on our sketches, though definite thrust figures seem to elude Frank [Halford] at the moment, though his weights and dimensions are on the mark. I have written to Corlett who is going to write to you directly, though he is confident there will be no cost to us for the alloys. Should he include his stress calcs, I'm sure you'll pass them on![73]

Cobb, for his part, seems to have waited for Ewan Corlett's letter before responding, while in Leningrad on business:

> Please find enclosed the dope from Ewan Corlett, I am now sure you are in your element!
>
> George [Eyston] was quite taken aback by your drawings, suggesting it is as far from a boat as it is possible to get, though I assured him it

TENSIONS RISE

> TELEPHONE: MANSION HOUSE 5544 (5 LINES)
> " " 4801 (3 LINES)
> " " 0306
>
> 15, ARTHUR STREET,
> LONDON, E.C.4.
>
> 1st July, 1949.
>
> Mr. Reid A. Railton,
> 2808 Oak Knoll Terrace,
> Berkeley,
> California.
>
> Dear Reid,
>
> Just a line to tell you that I asked George round for a drink last night and touched very lightly on the boat question. He seemed to think that Thornycrofts would be interested, and if anything is done in that direction, apparently Jack Thornycroft is the man with the "say so"; I asked him to keep it very confidential.
>
> He said he will be in New York on the 14th and 15th of August and will be very pleased if you would look him up on your way to catching the "Nieuw Amsterdam". I told him I would let you know and that you would very probably call him.
>
> I am going to Leningrad on the 21st July, and shall be back here about the 2nd or 3rd August. I look forward to seeing you shortly afterwards.
>
> With love to all the family.
>
> Yours,
>
> John

This letter from John Cobb, sent from his London apartment, shows that his discussions with Reid Railton about construction of a boat for the world water speed record were very well advanced as early as July 1949. *Cobb/Railton archives*

will float, so therefore, must be a boat. He has suggested a model to take to Thornycrofts, though Du Cane now seems to want to play the game. I shall be travelling back tomorrow, and look forward to seeing you on your arrival.[74]

Railton arrived in Britain on 22nd August 1949, armed with his own research data, and with concerns that Vosper was still clinging to what it knew best — conventional hydroplanes — and to that end was continuing down the *cul de sac* of the *Bluebird* jet conversion.

CRUSADER

From Cobb's company diaries, it is clear that he had several meetings with Railton, with George Eyston in attendance, but no note was made of any other participants. That Eyston was present suggests that these meetings were about Cobb's jet boat project.

Unknown to many in Cobb's project was that during October 1949 Railton was calculating the likely acceleration and velocity figures that the proposed boat could achieve with use of a Walter 509A rocket motor, as developed in Germany during the latter years of the Second World War, at first for missiles (including the V1 'Doodlebug') and then aircraft (notably the Messerschmitt Me 263 '*Komet*' fighter). After initial approaches to obtain a De Havilland jet engine, contacts in government offered the use of a requisitioned example of the compact Walter rocket motor, which could produce 3,300lb of thrust and weighed only 365lb. The fuel was a two-part mix of *C-Stoff* (30% hydrazine, 57% methanol and 13% water) and *T-Stoff* (hydrogen peroxide), which was extraordinarily unstable and could dissolve human flesh. Although use of this rocket motor would have alleviated problems such as excessive engine circumference and the need for air intakes, neither Railton nor Cobb was keen on embracing Nazi technology. Apart from the obvious objection, there was a known difficulty in 'throttling' the engine — as used in the Me 263 the throttle had just four settings — and some of the fuel ingredients would have been tricky to obtain.

In October 1950, a year after Railton's first investigation, Peter Du Cane was positive about the prospect, as outlined in this letter to Railton:

> Regarding the rocket motor project, this has so far progressed well as I have obtained the interest and co-operation of the Farnborough Rocket Experimental Establishment, who are keen as mustard and have already obtained provisional approval from the Ministry of Supply for co-operation.
>
> Handled properly I do not believe there will be much, if any, money involved and even the fuel I think might be wangled, though it is not so very expensive. The hydrogen peroxide is given as approximately four times the price of petrol, while kerosene, the other component, is more or less the normal type of turbine fuel.
>
> They can give us 3,500 lbs. of thrust for a total weight of 1,500 lbs., which is very good.
>
> What is better still is that the cross section is very small indeed and

TENSIONS RISE

15, ARTHUR STREET,
LONDON, E.C.4.

26.10.49.

Dear Reid

(Reakes) Many thanks for your letter which makes very satisfactory reading to me. As soon as you departed the weather broke and we have had rain and gales ever since. Godfrey Davis's bill came to £96, less than I thought, so I have debited you with £50 to keep to round figures and paid the rest myself.

Am glad you liked the C.E. I saw in the papers you were running late but as you say, they always slow down rather than take a bashing, a sentiment I entirely agree with at any rate when I am aboard.

George called on me a few days ago and I said I would let him know if we got going. I told him about the "habit of life" and non residents Income Tax and he went as white as a sheet! I hope to hear from you again before very long

All the best
John

P.S. No mail for you since you left.

In this handwritten letter, John Cobb tells Reid Railton that he would rather be slow and safe than proceed too quickly. Cobb/Railton archives

105

CRUSADER

A DISTINGUISHED HOTEL

The Whittier

Burns Drive · Detroit 14, · Michigan

November 5th. 1949

Dear John;

    Many thanks for your letter. I am back again at the old routine, and the Firm still seems glad to have me around. I expect I shall here until about the middle of December.

    I have been seeing a good deal of Douglas Van Patten -- the Miss Canada designer. There seems to be no doubt that the boat has been officially timed one way at over 150 mph, but the bronze propeller has got bent with the effort every time, and prevented a satisfactory return run. The sister-ship, Astraea, is here in Detroit, and I have been having a good look at it---also taken some excellent stereo-photographs of the shape, which may come in useful some day.

    I have come to the conclusion that the bottom-formation of these boats is pretty close to what we want, but I dont want to confuse Peter by urging it on him at the present stage. After all, he might turn up something just as good or better.

    Thanks for your generous settlement with Godfrey Davis. I think it was a very satisfactory solution of the transport problem, and I should certainly do it again.

    It snowed hard yesterday for an hour or two, but the ground was too warm for it to lie. However, it was doubtless a foretaste of hell to come.

    Yours ever,

A letter from Reid Railton, written from his Detroit hotel, The Whittier, discusses the revolutionary hull design of the Douglas Van Patten-designed *Miss Canada IV*.
Railton/Cobb archives

will enable our aerodynamic problem to be simplified.

The rocket propulsion scheme is, of course, subject to it being proved safe, controllable and practicable, but with the sympathy of the Research departments on our side they do not consider this to be a big problem. They point out that in America a considerable number of bi-fuel rocket fighters are flying now, and of course they were in Germany at the end of the war.

I think it desirable to keep this method of propulsion strictly to ourselves; as if successful it is so very easily adopted by other contestants.[75]

Later, however, Du Cane wrote: "Fitting a rocket to such a machine seems more appropriate for the day after tomorrow, than for serious record breaking today, but perhaps I am too old fashioned?"[76]

Returning to 1949, Railton was back in America by 21st October, when he wrote to Cobb with a summary of his conclusions from his spell in Britain:

As ever, many thanks for your efforts and hospitality, though I am sure the weather was not of your doing.

It is gratifying to see how keen George is, and feel that whoever finally builds the boat, it will have been his enthusiasm that led to it, and he will be invaluable as the diplomatic go between.

The Apel formula pretty much ties us on position and dimensions between the planing surfaces, but there is nothing that says we cannot turn the thing around, which I think George now sees. Whether we can get Du Cane or anyone else to see it is of course another matter. Frank [Halford] is a good man, though the size of the Ghost makes for a tall hull, but fits length wise. Whereas an axial flow, gives us a suitable height, but moves the cockpit too far forward. Rolls-Royce do not seem keen, and Frank has De Havilland ready to go, so it appears it would be a modified Goblin or a Ghost. Fortunately, the stern is the wide end![77]

And so the trans-Atlantic communications resumed. Cobb replied on 26th October:

Many thanks for your letter which makes very satisfactory reading to me.

CRUSADER

> As soon as you departed the weather broke and we had rain and gales ever since!
>
> George called on me a few days ago and I said I would let him know if we got going.[78]

Railton's reply, dated 5th November, came from his Detroit 'address', The Whittier Hotel, where he based himself while working for the Hudson car company.

> Many thanks for your letter. I am back again at the old routine, and the firm still seems glad to have me around. I shall be here until about the middle of December.
>
> I have been seeing a good deal of Douglas Van Patten, the *Miss Canada* designer. There seems to be no doubt that the boat has been officially timed one way at over 150 mph, but the bronze propeller has got bent with the effort every time, and prevented a satisfactory return run. The sister ship, *Astraea*, is here in Detroit, and I have been having a good look at it, and also taken some excellent stereo-photographs of the shape, which may come in useful some day.
>
> I have come to the conclusion that the bottom formation of these boats is pretty close to what we want, but I don't want to confuse Peter [Du Cane] by urging it on him at the present stage. After all, he may turn up something just as good or better.[79]

Before replying to Railton, Cobb felt the need to try once again to get some firm answers from Peter Du Cane. His next letter to Railton seems to show that he was unhappy with what he learned during a visit to Vosper:

> Many thanks for yours of 5th Nov, and I am not surprised to hear that they still want you around. I went down to see Du Cane last week as I had not heard from him and was very surprised to learn that in spite of the overwhelming evidence you put before him, he had insisted in making a model with the floats fully attached to the main hull. It turned over, fortunately not at Haslar, and he has now mounted the floats on outriggers but has not been able to run owing to trouble with the Jetex [rocket] charges. I sincerely hope you will be able to come over next Feb, my central heating is working a treat!

TENSIONS RISE

I hope this will catch you before you go home, anyhow all the best for Christmas and 'Noo Year'.[80]

This update obviously concerned Railton enough for him to cable Cobb at his 'Cobang' telegram/cable address at Anning, Chadwick & Kiver on 5th December: "Will write to PDC. Informed him weeks ago fully attached floats were a dead end. Has he done nothing with my drawings?"[81]

The next day Railton wrote a fuller response to Cobb:

I have just received two letters from Peter Du Cane, both rather vague and rambling, to which I have sent the enclosed reply. It all sounds a little woolly to me.

In one of his letters Peter appeared quite anxious for concrete suggestions, and wanted to know how we should stand with Van Patten if it seemed expedient to experiment along his lines. Hence the last paragraph of my letter to him.

However I think it's about time you had an interview with Peter and emphasised the importance of getting, before the end of January, some definite results which will indicate either (1) that his present line of investigation is promising and appears only to need a course of trial and error in the tank to get the results we want, or (2) that our present line is unpromising, and that a quite new approach is needed.[82]

It is not difficult to deduce what that new approach would be. Between them, Railton and Van Patten had been getting some startling results, both in low-speed transition to planing and encouragingly little drag resistance at speed.

One of the author's interviewees who knew Du Cane, but did not want to be identified, thought he "seemed to be floundering a bit, the feeling being that he had other things on his mind and was not concentrating on the project properly."[83]

On the '100% Van Patten' model, the triple-chine underside provided for a very low-speed breakaway but still produced quite high running-resistance figures. The Railton/Van Patten hybrid was slightly slower to get onto plane, because of the width of the forward low-speed step, but once up on the running surfaces it produced a very flat speed/resistance curve. It would appear that all through this period Railton was feeding Vosper with data, and drawings, but that the information was, for some reason, not being acted upon in its entirety. Vosper's models began to look very similar to Railton's in all but detail.

This was not lost on Cobb, who wrote back to Railton on 29th December 1949.

> I received your letter dated the 6th December, and copy of yours to Peter. Very shortly afterwards, he called on me at the flat, and it appears that he had a satisfactory run with the new model but cannot get it 'over the hump' from rest yet, and has been running it down a ramp to launch it. He still seems to be in trouble with the rockets, and I can't understand why they have not yet been put right, and it makes the experiments a bit dubious. I was glad to note from what he had to say that he was quite keen on the Van Patten idea, and personally I am flat out for it unless he produces the answers within the next few weeks, which I doubt if he can do on his own, and I think privately he would be very glad of the assistance. He also said that he would want a lot of help from you on the construction when it came along, and as obviously you cannot afford the time to hang around Portsmouth indefinitely, it might be that Van Patten might be able to lend a hand as well in this direction if you thought it advisable. However, I hope Peter will have some more results by the time you return to Detroit, and I feel that this will be the time when we must make a decision.[84]

Railton's reply, dated 4th January 1950, was hand-written on a Hudson Cars compliments slip, marked 'Private and Confidential', and — unusually — sent to Cobb's Surrey home:

> Words fail me. I have given our data to Peter on several occasions, and fail to see why they cannot get their model to un-stick, they should be much further along, there seems to be a resistance to what we have achieved over here, or *how* we have achieved it. I feel, there is the rub. We need to get Frank to confirm we have the engine, sooner rather than later, and we need to ensure Peter starts working from not only his, but our data and drawings as well. We are wasting so much time with these dead ends.[85]

Cobb felt he was firmly stuck in the middle, at the centre of things in Britain,

TENSIONS RISE

TELEPHONE: MANSION HOUSE 5544 (5 LINES)
" " 4801 (5 LINES)
" " 0306

15, ARTHUR STREET,
LONDON, E.C.4.

29th December, 1949.

Mr. Reid A. Railton,
2808 Oak Knoll Terrace,
Berkeley 5,
California, U.S.A.

Dear Reid,

    First of all, thank you very much for the 3 bottles of Scotch which arrived safely just before Christmas; it was very kind of you to think of me and I appreciate it immensely.

    I received your letter dated the 6th December, and copy of yours to Peter. Very shortly afterwards he called on me at the flat, and it appears that he had a satisfactory run with the new model but cannot get it "over the hump" from rest yet, and he has been running it down a ramp to launch it. He still seems to be in trouble with the rockets, and I can't understand why they have not been put right, and it makes his experiments a bit dubious. I was glad to note from what he had to say that he was quite keen on the Van Patten idea, and personally I am flat out for it unless he produces the answer within the next few weeks, which I doubt if he can do on his own, and I think privately he would be very glad of assistance. He also said that he would want a lot of help from you on the construction when it came along, and as obviously you cannot afford the time to hang around Portsmouth indefinitely, it might be that Van Patten might be able to lend a hand as well in this direction if you thought it advisable. However, I hope Peter will have some more results by the time you return to Detroit, and I feel that this will be the time when we must make a decision. In any case I imagine you will not be seeing Van Patten until that date.

    Peter had a very unsatisfactory talk with Halford some time ago and since then he has been investigating the rocket proposition with the people at Farnborough who, he says, are very keen indeed on the idea. Here again I am afraid I shall want to lean on you heavily for advice and it is a matter which won't wait indefinitely.

p.t.o.

Here Cobb starts to show his frustration with Vosper's lack of progress with model testing and concurs with Railton that perhaps Douglas Van Patten could take a bigger role. Cobb/Railton archives

CRUSADER

15, ARTHUR STREET,

LONDON, E.C.4.

19th January, 1950.

Mr. Reid Railton,
Whittier Hotel,
Detroit, Michigan,
U.S.A.

Dear Reid,

I received your short note from Berkeley, and presume you are now back at work. No doubt you have received my previous letter, but I am finding it very difficult to know what to do.

I saw Peter last Tuesday, and really he has got no data much to go on. He blames the rocket charges for the delay. Apparently they cannot get these quite reliable. He has been making trials but has not yet managed to get the model to go over the hump under its own fuel. He says the nose will not rise. Altogether I find him terribly vague, but he seems very keen on getting drawings from you for a new model for the Van Patten hull; so whether it is worth your while to come over now I find it hard to say, except, of course, your presence would no doubt stimulate him to further efforts. I have no doubt you could probably put him on the right lines.

Also, there is the question of deciding what form of power plant we are going to use, and in this connection we spent a day at the R.A.E. Rocket Establishment near Aylesbury last Tuesday, which was very interesting and, of course, Peter is all out for this method, but to my lay mind there appears to be a great many 'ifs and buts' about this proposition, not the least of which is the fact that fuel for a minute at present rates costs £100, but this might be overcome, as they are definitely very keen. Also they have not yet got a motor which could be controlled by a throttle, but they say they could produce this modification within six months.

Personally, I would be very happy to see you over here on the basis of our original discussion, but would like to

p.t.o.

Frustration grows: in this letter Cobb states that he finds Vosper's Peter Du Cane "terribly vague". Both Cobb and Railton felt they were not receiving anywhere near enough data from Vosper, something that Railton would have found extraordinarily irksome, especially as he was so generous in his assistance to others. At times it appeared as if they were fighting to learn anything constructive. *Cobb/Railton archives*

TENSIONS RISE

unlike Railton, but by his nature unable and unwilling to upset anyone or anything. This is the letter he wrote to Railton on 19th January:

> I received your short note from Berkeley, and presume you are now back at work. No doubt you have received my previous letter, but I am finding it very difficult to know what to do.
>
> I saw Peter last Tuesday, and really he has got not much data to go on. He blames the rocket charges for the delay. Apparently they cannot get these quite reliable. He has been making trials but has not yet managed to get the model to go over the hump under its own power. He says the nose will not rise. Altogether I find him terribly vague, but he seems very keen on getting drawings from you for a new model...[86]

That same day, Railton wrote to Du Cane:

> I find that I still have to acknowledge your letters of Dec 12th, Jan 3rd and Jan 4th. I'm afraid that this is rather remiss, but I have really been waiting to receive the photographs you mentioned in the first letter. Were these ever sent, for they certainly haven't arrived? [A margin note by Jean Carpenter says, 'Sent on 6/12/49 with copy of letter dated 28/11/49 to Berkeley (California) by surface mail, hope they are not lost.] I opened an exciting package from you about Christmas time, but it turned out to be a very fine desk calendar, for which many thanks.
>
> I'm sorry your experiments have met with so much difficulty, both with the rocket units and with behaviour of the model. I don't think I can make any useful comment at this distance, particularly as I don't even know what sort of planing surface you are using, i.e. whether plain skids or a more conventional hydroplane shape. There are hundreds of questions I should like to ask, but I realise it would just be a waste of time and stationery.
>
> I haven't decided anything about making a trip over, and shall not do so until I have heard again from John. I have arranged my work here so that I could come over towards the end of February for a week or two, but later than that would be almost impossible until the end of June when I intend to come in any case.
>
> Nothing further has been done with Van Patten. However, I now think that the only sensible way of using him would be to take him

completely into our confidence, and have him collaborate with me here in getting out a set of lines. It might even be worth considering doing some preliminary model tests here if we could get the facilities. In any case I can't make any firm plans until I get a definite lead from you or John.

I was most interested to hear about your negotiations with the rocket people at Farnborough, and agree that they sound most promising. My own feeling is that a thrust of 3,500 lbs is on the small side, and would necessitate a boat of smaller dimensions than we have so far envisaged, and that this might be a handicap from the point of view of ultimate speed, although it would certainly be less expensive to build. Of course, if De Havilland and Rolls-Royce both refuse to play, your rocket idea is the only alternative and an excellent one at that, though I think it would make 200 mph rather problematical.

I have ordered your book on rockets and will send it along as soon as it arrives — and after I have read it myself![87]

This letter clearly shows the lengths Railton and Cobb had to go to in order to drive the project along, and in the right direction. Both had already looked into the use of a rocket motor and dismissed it, as is evident from Railton's prior knowledge that the unit could not achieve 3,500lb of thrust, and of course Railton was already testing hull forms and sending the data to Vosper.

As Railton's letter was flying over the Atlantic towards Britain, Cobb's arrived in America. Whereas Cobb, though obviously frustrated at the lack of progress at Vosper, was being diplomatic, the contents of his letter clearly irked Railton, as is evident from this short letter written by Railton on 28th January: "After much thought I have come to the conclusion that it is 'my move'. Accordingly, I had Van Patten over here for the night last week, and have discussed everything with him quite frankly, except that I haven't mentioned Peter's investigations about rockets, which hardly affect the present problem anyway."[88] This was followed very quickly by a cable message: "Did not discuss T&T scheme with DVP. Keeping powder dry."[89]

At the same time, Railton wrote again to Du Cane, obviously feeling that his previous subtlety was now misplaced:

> I enclose herewith a letter I have just written to John and which you will find self-explanatory. I gather from both your letters and from

TENSIONS RISE

Douglas Van Patten was as happy on a boat as designing one. Like Railton, he was capable of seeing unexpected solutions to problems that beset projects he was working on. *Private collection*

what I have heard from John that you will welcome this alternative line of approach.

In order that we may learn as much as possible from what you have done up to now — even lessons of what <u>not</u> to do — I wish you would send me a sketch or photograph of the models you have tested, showing (1) the bottom formation, and (2) the C.G. position. At present I have no knowledge of this.

As I remember, we decided on a 1/16th scale as being suitable for self-propelled experiments with available Jetex units. Will you confirm this? My idea is to stick to this scale so that any promising models made here can be shipped to you for test with the Jetex equipment.

I am rather coming round to the thought that the actual tests for the getting over the hump condition could be done most conveniently in a small slow-speed tank, if we could find one. The dollar situation may make it very difficult to do this here, and one rather shudders at the thought of using the Haslar sledge-hammer to crack such a small nut. On the other hand, the Jetex method seems to have drawbacks which you have so painfully discovered, and may prove to be further complicated by requiring solution of directional stability problems that we don't want to be bothered with at the moment.

I have been toying with the old idea of towing the model with a speed-boat, but in a rather special way. I imagine that the chief difficulty is to maintain a constant and stable line of thrust, and it has occurred to me that this might be done as in the accompanying sketch. The idea is to push a small planing pontoon ahead of the boat, and to push the model with a bridle from the pontoon. What do you think? Obviously it wouldn't exactly reproduce the action of the jet, but it should be near enough for these 'hump' trials.

I will of course keep you posted on any progress we may make. I am sending John a copy of this letter.[90]

This letter would have crossed with one from Du Cane to Railton, dated 30th January:

I hope by this time you will have the photos of the model. Not that this will do a great deal to help us here, as we have, of course, modified it a few times since then to endeavour to get it to 'unstick'.

As I told you it runs O.K. when launched from the modified ramp.

It is as you say impossible for you to do much about it at that distance, though I have learned a lot and would like to tell you of it. Admittedly some is what we cannot do rather than what we can! Equally I am rather at a loss to know what to say about the Van Patten question.

It is a little difficult to see how he gets over the aerodynamic problem as from photos I have seen the hull would seem better than the Apel type in this respect, but by no means perfect.

There is also the question as to what happens between 150 - 200 M.P.H. as regards behaviour. Can you model test to these speeds without U.S. Government assistance?

We here have our ideas on how to interpolate but there are so very many factors coming into it, so must obviously reserve judgement before knowing a bit more.

I think the idea of trying the principle is sound and good, but personally am against giving our ideas on rocket model testing, spoilers, etc. etc. away. Also it is a pity to hand over the idea of using the bi-fuel rocket, at least without knowing more of what Van Patten has to offer.

TENSIONS RISE

Can it be arranged that Van Patten puts forward his idea of the solution to produce a boat for 200 m.p.h. with displacement approximately 6,000 lbs, diameter of GHOST for dimensions. When he has done his best and produced lines or model we could test it against the best we can do here and judge finally on merit.

In any case it will in all probability require modifications here in the final stages.

It is all fearfully difficult to explain and I wish to goodness there was not 5,000 miles or so between us.

This does not mean that we do not see Van Patten's or your point of view, so I suppose the only thing to do is carry on here to the best of our ability, which we are doing, until the matter over there is clarified.

Gawn is going to run the model(s) in the tank shortly pending arrival of the new type rocket, so we get a good idea of drag and behaviour, also thrust required to unstick.

John suggests you come over the middle or end of February to discuss the general position. Can you then bring some dope on the Van Patten idea so that we can get a better idea of what it is all about? Also there is the power unit question to settle.

As an alternative what about my flying over in the very near future for a day or two to discuss the point and save you coming over till Van Patten has his model ready or till June. Perhaps after a preliminary talk with you I could see Van Patten with you?

If John approves.

As matters stand at present am supposed to be off on a trip in a yacht job early March.

If you would like me to come over give me a cable on receipt as must organise.

Gawn's speed/resistance curves for the GOBLIN 'BLUEBIRD' indicated 3,000 lbs for 200 M.P.H. so do not see why 3,500 lbs should be far out even if drag is increased.[91]

Cobb's response came on 30th January. In it he informed Railton that he had visited Vosper to discuss the current state of development with Du Cane. Cobb told Railton that Du Cane was just about to send him a letter, and stated: "You will note from the letter, he is a bit hesitant about Van Patten, although he would like to have his ideas to try out! However, after we had been talking

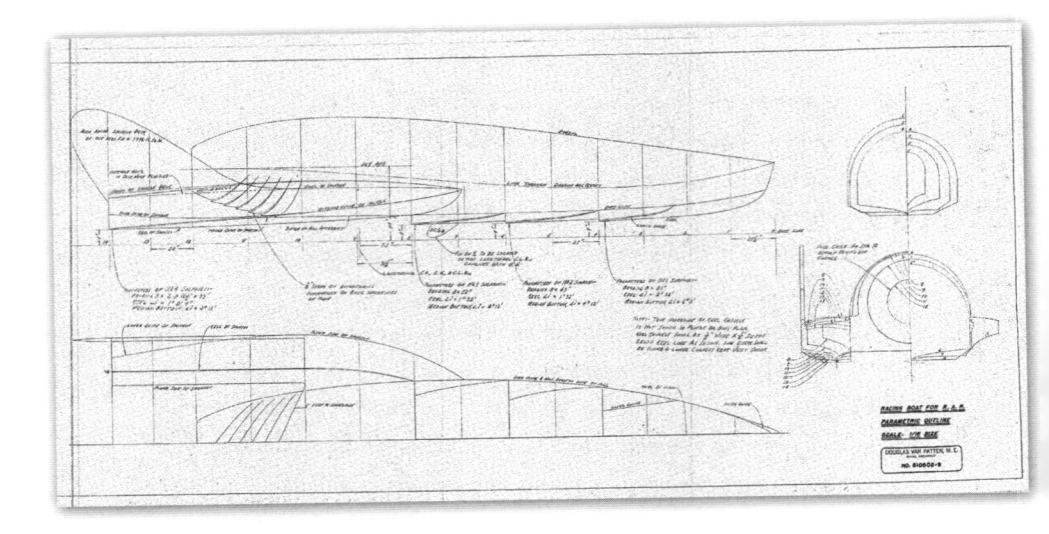

The Van Patten/Railton design displays a striking similarity to the general side view of *Crusader*. This is the blueprint sent to Vosper; note that all of the planing surfaces incorporate a vee profile. *Railton/Van Patten archives*

for some time he suddenly produced the idea that he should take a plane and make a quick trip to see you in Detroit. I have no idea of your other business, and have prepared the way for you to give a negative reply to the idea. I should add, Peter thought he should travel at my expense!"[92]

Again, Railton's reply, dated 2nd February, gives a clear indication that his own scheme, and that of Van Patten, though similar, were in advance of those being tested by Vosper:

"My reasons for leaning towards a three-step design make rather a long story. Fundamentally it is because I think I see a way of combining satisfactory aerodynamic characteristics with a practicable form of bottom. Actually I have provided Van P. with a dimensioned sketch of what I have, and have asked him to work up the nearest approximation to this that he thinks will be satisfactory hydro-dynamically. I am having another meeting with him tomorrow."[93]

It is fairly obvious that Railton was looking at the big picture, while Vosper and Du Cane were fairly fixed in their thinking, and struggling with it. That is not to say that Railton was not open to other ideas, as he often stated. Vosper, or Van Patten, might separately come up with certain data or suggestions that solved problems they all might have. As for the scheme arrived at with the help of Reg Beauchamp and Ewan Corlett, Railton was continuing to keep this back, as he still felt it was *so* radical.

TENSIONS RISE

In the meantime, Railton received another letter from Du Cane, dated 6th February and copied to Cobb:

> Have now received your two letters of January 28th and February 2nd (addressed to John).
>
> The trouble has, of course, been that in endeavouring to tackle the problem systematically I have been severely held up by failure of the rockets and have at last decided to go in for intensive model testing on more orthodox lines. It has probably been bad judgement on my part that this decision was not arrived at previously.
>
> Am enclosing drawing showing model tested up to date in case the photographs have gone amiss. This also shows the line of thrust and C.G. position. These could be altered to a certain extent if any advantage could be obtained thereby.
>
> The modifications so far have consisted in endeavouring to give the nose or bow more hydrodynamic lift in order to assist in unsticking. One, of course, has to be careful here not to make her dangerous aerodynamically.
>
> There is no question that drastic alteration would remedy the situation but have been tackling it stage by stage. In small doses.
>
> Scale model 1/16th should be O.K., while as regards the line of thrust, I do not think in the first instance you need worry so much

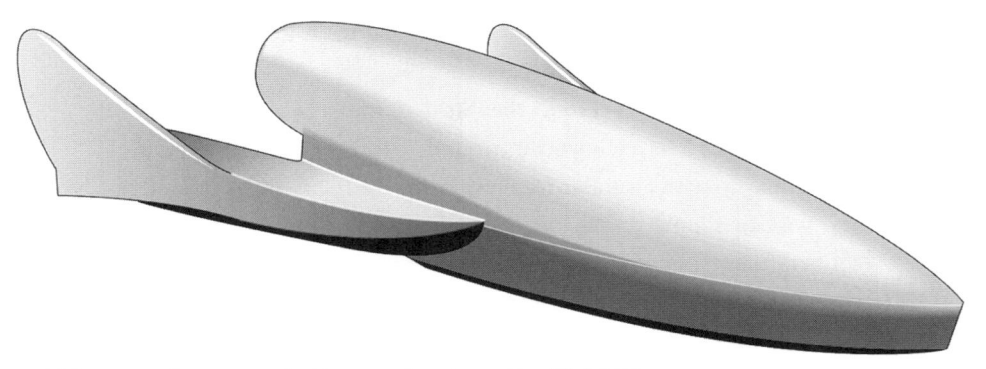

This computer-generated image, developed by Mick Hill from the original drawings, shows how the Railton/Van Patten *Crusader* might have looked. Later drawings showed a forward planing surface more like the finished version, although of a double chine layout, but the main hull section here is unmistakably *Crusader*. Mick Hill about pushing the model provided you have reproduced the actual line of thrust. Certainly you should be able to tow on an outrigger, clear of the wake.

I also have some ideas on another form and will play about with them on my own, but in the meantime expect we shall be seeing you over here in June, if not before.

Good luck to you.

P.S. Am by no means certain as yet that the present line of investigation is not perfectly sound and capable of yielding the result we want.[94]

Clearly, Du Cane was struggling, as he was still introducing 'new forms' to test, whereas Railton was honing his original idea. Railton's next letter also demonstrates how far ahead he was in his own design work. Vosper's archive shows that the company realised only in June 1951 that there was a need for the rear sponsons and/or their supports to be removable, for transportation, as confirmed at that time by Du Cane in a letter that included the construction plans. Railton, on the other hand, had addressed this issue 16 months earlier, as confirmed in his reply directly to Cobb, dated 11th February 1950:

Will you find out and let me know by airmail what in England is the maximum width of a load that can be legally carried on a truck on the highway, also what is the maximum that the railways will accept? It has occurred to me that it would be a great advantage to keep within these dimensions with the boat. It would be most annoying to find ourselves a few inches out through ignorance.

P.S. [hand-written] From the Apel formula for length/width, regardless that I have flipped it around, I fear we will be over, so I shall come up with a similar framework for the floats as the main hull, and we can simply detach them, but I would be grateful if you can confirm the maximum width with Henry Spurrier. If Peter comes up with something different, his ideas may be narrower, but the formula says they shouldn't be by much![95]

Railton's concerns were confirmed in Cobb's next letter of 17th February:

I received your letter of the 11th, and for your information, the maximum width of a load on the highway in this country is 10 feet

9 inches. The railway dimensions are not quite so precise as they vary a little with the height of the load, but in this instance I think 9 feet would be the figure.

I haven't heard anything more from Peter, except I think he is redoubling his efforts now that he knows he is in competition, which is all to the good I think.[96]

On 5th March, Railton sent Cobb what he called an 'interim' report. He confirmed that the models he had been testing were all making the transition from rest to the planing condition at quite low speeds, but that the Van Patten/Railton hybrid was showing some "quite lofty high-speed resistance figures". He was, however, quite pleased to inform his friend that the model of his preliminary scheme was "up on its points and then a slippery little devil, with some promisingly low resistance numbers. As far as Peter is concerned, I think he needs the purebred Van P. model sooner rather than later to combine the best of both, he then might make some progress. The tank here is too slow for the data to be of great help, they simply won't compare to that at Haslar, so for the time being I have enclosed a copy of Van P's blueprints."[97]

Cobb replied from his Arthur Street address on 8th March.

I have to acknowledge receipt of your letter of the 24th February, and also one received this morning enclosing the blue print.

It is very heartening to me to see that you are making such progress, as I was getting somewhat depressed with the way Peter was failing to get going. I do not think there is any doubt that the fact that you are now taking a hand in the proceedings has stimulated him to put his head down, and I had a short note from him about a week ago informing me that he has now got the rockets satisfactory and has made some encouraging tests; it did not seem necessary to him to tell me what they were, and as you say, at the end of his letter he announced his intentions of going off for two weeks holiday however! I will send the blue print on to him when he has returned. From the look of it, it appears that you have brought the sponsons in close to the main hull, but I will not ask you a lot of questions now, as I shall no doubt be hearing further from you, and in any case, will spend a few dollars phoning you when I get to New York. I take it there is no intention of using air rudders even if you do decide to use the fins.[98]

CRUSADER

Airmail.

15, ARTHUR STREET,

LONDON, E.C.4.

TELEPHONE: MANSION HOUSE 5544 (5 LINES)
" " 4801 (5 LINES)
" " 0306

8th March, 1950.

Mr. Reid A. Railton,
Whittier Hotel,
Detroit, Michigan.

Dear Reid,

I have to acknowledge receipt of your letter of the 24th February, and also one received this morning enclosing the blue print.

It is very heartening to me to see that you are making such progress, as I was getting somewhat depressed with the way Peter was failing to get going. I do not think there is any doubt that the fact that you are now taking a hand in the proceedings has stimulated him to put his head down, and I had a short note from him about a week ago informing me that he has now got the rockets satisfactory and has made some encouraging tests; it did not seem to occur to him to tell me what they were, and as you say, at the end of his letter he announced his intentions of going off for two weeks holiday; however! I will send the blue print on to him when he has returned. From the look of it, it appears that you have brought the sponsons in close to the main hull, but I will not ask you a lot of questions now, as I shall no doubt be hearing further from you, and in any case will spend a few dollars phoning you when I get to New York. I take it there is no intention to use air rudders even if you do decide to use the fins.

I saw Wooding yesterday and he told me he has already mailed you the draft of his proposals, so he has kept his word.

I had a letter from Audrey the other day and she enumerated the programme that she, you and Tim proposed to carry out over here. As you know, my flat is entirely at your disposal the whole time you are here, and I do not see any reason why we should not tuck Tim into my Dressing Room if he is going to be around Town some of the time; I can assure you it would be no trouble.

p.t.o.

More frustration: writing of Peter Du Cane, Cobb stated that he felt Railton's involvement had "stimulated him to put his head down". The relevant archives do not contain anywhere near as much correspondence from Vosper as from Railton and Cobb. *Cobb/Railton archives*

TENSIONS RISE

Meanwhile, Railton received a letter from Du Cane dated 7th March and not copied to Cobb:

> Thank you very much for sending me the book on rocket propulsion.
> I am just off for about 10 days in a yacht down to Gibraltar and shall be flying back.
> As regards the model, things have gone very much better lately; I have at last been able to get the I.C.I. rockets working and we have done some extensive testing over the last ten days. The resulting model runs satisfactorily subject to a little 'trimming' here and there, but have sent it over to the Haslar tank for detailed performance estimates.
> Though the main body still remains, I have had to do some fairly substantial modifications in other directions and it will be desirable to check that the aerodynamic qualities have not been upset. This I can arrange later.
> I am making arrangements to check at least one more model which I think may have possibilities.
> More details later.[99]

Railton's response on 27th March was short, and to the point:

> I enclose a copy of a letter I am writing to John.
> I expect he has sent on to you the print I sent him recently showing the scheme in question. I wish you would send me a line to Berkeley giving me the latest dope, and also a sketch or print or something showing the lines you are working on. It would be of interest to see how near or far apart we have got on our separate lines of investigation.
> Looking forward to renewing our consultations at the end of June.[100]

The same day, Railton then wrote a long letter to Cobb that contains some interesting snippets:

> We have just completed a tank test at the University of Michigan which takes me about as far as I can go over here.
> Van Patten had an excellent 1/16th scale model made of this design which we have cooked up between us, and the University, at his entreaty, was kind enough to test it free of charge. They managed to

tune up the carriage to travel at practically 10 mph, a record for the course, which gave us a scale speed of 40 mph. This was enough to take it well over the hump, but of course I have no idea yet what the high-speed resistance will be.

The question now arises as what to do next, and my suggestions follow. I think this model should now go to Peter for tests at Haslar, both as regards resistance at higher speeds, and also for behaviour (porpoising, etc). If it still appears promising by these standards and with what Peter has been doing himself, I think that a new wind tunnel model should be made (I will supply the drawings) and it should be ready for test by the time I get to England. In other words, the Haslar tests should be completed in time for the wind tunnel model to be ready by that date. I could then verbally inform Fairey's of the exact data we should want on it.

In order to keep him informed, I am sending Peter a copy of this letter, down to the pencil mark. You might encourage him strongly to send me real information of what he is doing. This has been conspicuously absent so far, though I strongly suspect there has really been nothing to report!

P.S. [hand-written] I received a missive from Peter about the rocket book, which in his mind is the solution to all his problems, but my guess is he won't read it. He also says he has yet another model to test which begs the question if he has included ANY of the data that we have sent to him.[101]

The 'PS' is particularly interesting and, of course, it is abundantly clear that both Cobb and Railton were becoming increasingly frustrated with Vosper's lack of progress. It is also evident that Railton was still keeping his own model close to his chest.

Fairey is mentioned for the first time in this letter. Founded in 1915, the Fairey Aviation Company was one of the first aircraft manufacturers to construct its own wind tunnel, at its headquarters in Hayes, Middlesex. The company produced some iconic aircraft, not least the Swordfish, one of Britain's last military biplanes. The Swordfish's ability to take off and land on an aircraft carrier, and carry torpedoes capable of damaging German warships such as the *Bismarck* and the *Tirpitz*, was down to the high-lift/low-drag shape of its wings. One of the aerodynamicists responsible for this part of the design

TENSIONS RISE

was Maurice Shadwell Hooper, who became the engineer in charge of Fairey's wind tunnel post-war. For Railton, he was the 'go-to' man.

At this point it is instructive to return to the role of Thomson & Taylor at Brooklands, as recorded by Reg Beauchamp: "We were not heavily involved by now, we'd done some drawings, and calcs but Reid kept us informed, just in case, he was from the 'belt and braces' school of thought. We received some data from some tests he had done in Michigan, comparing two models. He sent two sets of drawings, one Van Patten, and the other one I recognised from the discussions we had had in his office at Brooklands. He asked that we come up with a neutral aerodynamic shape for the upper hull of the Van P model, that borrowed heavily on our original while instructing us to get a model made from the T&T drawings but incorporating the mods he had drawn up to include part of Van Patten's multi-step layout at the bow, and to get it made to scale to test at Haslar."[102]

As Cobb arranged to pick up the Van Patten model, he wrote again to Railton, who was still quietly working on a 'Plan B'. This letter is dated 5th April:

> I received your letter of the 27th March a couple of days ago, and am very pleased to read about your progress. I have written to Van Patten, and look forward to seeing him in New York, and I can assure you that I will express my thanks to him to the best of my ability.
>
> I went down to see Peter yesterday, and he has had some quite encouraging tank tests, but I feel pretty certain nothing like as good as yours will be. He has promised to write you fully and enclose a picture. I stressed the point to write to Berkeley, but you know how vague he is, and I am more strongly of the opinion than ever that he is not capable of sustained coordinated thought.
>
> In my humble opinion, his model looks as though it would react unfavourably aerodynamically, but of course, I might be wrong. His second model looks to me exactly like the model we had in the wind tunnel without any sponsons, and, therefore, I should have thought unstable laterally. He endeavoured to give me a demonstration of this second model yesterday with the usual comic results. After capsizing, it leapt about like salmon on the end of a hook![103]

As Cobb was discussing a craft he may have ended up piloting, and at over

200mph, his description of the test as 'comic' is a superb illustration of his stoicism.

Railton's response simply reinforced Cobb's statements: "Welcome to the madhouse. Just had a letter from Peter, still very vague, but enclosing a photograph of his latest model. From the look of it I should say that he hasn't begun to consider things like reasonable safety, seaworthiness, structural matters etc, and other things Van P and I have been sweating blood over. However, we shall see."[104]

Cobb, now in New York, was able to write a quick response from his hotel, the Governor Clinton, just before his return to England. He confirmed that he had the Van Patten model and asked Railton: "No doubt you will send Peter drawings for the wind tunnel model with instructions? The model certainly looks peculiar, rather like a bumble bee's arse from behind!"[105]

Obviously, both Cobb and Railton had a firm belief in the concept they had come up with at Reid's Brooklands office before they had contacted Vosper, and now both men increasingly lacked confidence in what Vosper was attaining, and how, for the same reasons.

Another letter from Du Cane on 5th April still did little to keep Railton informed of what was happening at Vosper:

> Thank you for your letter of March 27th with enclosure to John. I have also received a print showing your developments up to date in conjunction with van Patten and am much interested thereby.
>
> It is really very difficult for me to make much in the way of criticism of this design, as much will depend on how she actually behaves in the course of some model tests. The figures you give appear very encouraging, but I am at a little bit of a loss to understand exactly what happens at the after end of the boat. We certainly ran into difficulties with something approaching this arrangement after you left, but the cases are not entirely parallel.
>
> As regard the air fins, I myself would strongly back water fins of reasonable dimensions as being worth a good deal more than anything in the air. Anyway you will be pleased to hear that I have arranged with [Richard] Gawn to test your model as soon as it comes over and it will be most interesting to see the results.
>
> Now as regards what I have been doing, I think the quickest thing will be to send you a photograph of the craft as at present developed

TENSIONS RISE

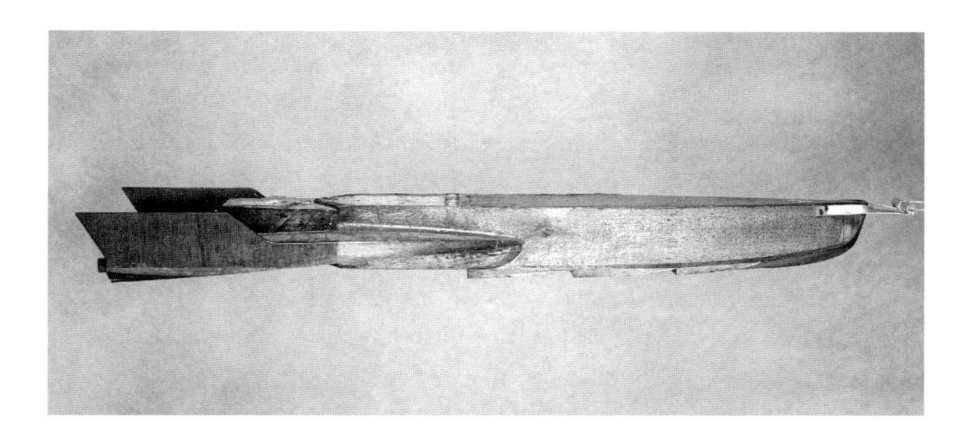

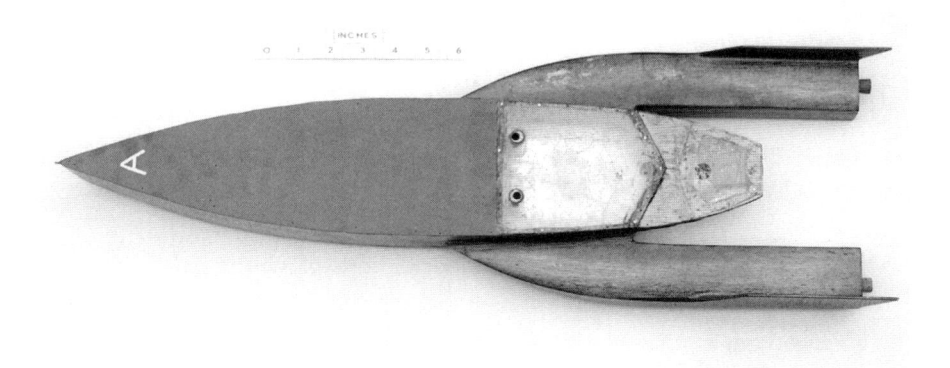

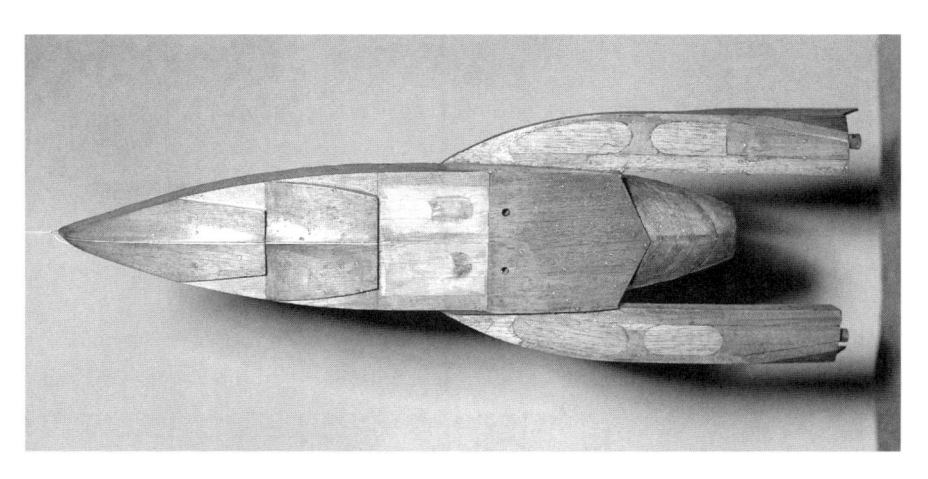

Views of the Van Patten model. Railton saw something in this unusual hull form, and Van Patten was generous enough to allow Railton to 'borrow' certain aspects for *Crusader*. Railton/Van Patten archives

during the first series of runs in the tank. There is a certain amount of 'grooming' required yet, but mainly minor details.

With 5,000 lbs available thrust (this at 5,800 lbs displacement much less if we can reduce displacement) there is little doubt at the moment that we can do the job, although from the point of view of resistance high up the scale there is a certain amount to be done, such as reduction of fin area and various planing area reduction processes which are relatively simple now that the behaviour is established to be very good.

You will undoubtedly see that the nose or bow has had to be altered in the interests of 'unsticking' and this will require further wind tunnelling, but I think it can be reduced and made neutral from an aerodynamic point of view. However, admittedly this hurdle has still to be finally settled.

This model as far as it has progressed fundamentally runs on the same surfaces as in the case of our original three-legged model, which is really the best of the lot; but of course has its drawbacks we all know about.

I am working on two other models, both of which have some promise and one of which has been suggested to me following visits to Farnborough.

There is no doubt about it that the latter will be extremely fast from the point of view of unstick and resistance but I am not too happy about its water behaviour high up the scale. However, it is early days and if this one comes to anything it has certain definite advantages from the point of view of simplification and therefore cost.

Hope this will give you something to go on and that by the time you come over here in June we shall be ready to make a final decision which I believe now to be quite on the map.[106]

Railton's initial response to Cobb was hand-written on a Hudson Cars compliments slip on 6th April: "John, a quick note. If you have received Peter's of the 5th, you'll be as maddened as I am! The after end is clearly marked 'truncated for jet exit', so why on earth is he at a loss? As for his obsession with increasing the size of the underwater fins, and therefore drag, is more than frustrating. We knew the surface positions an age ago, yet he is still referring back to the 'egg' [on stilts], and ANOTHER two models. I shall resend our

TENSIONS RISE

data, yet again! As for the photographed model, NO, no, and no!!! More anon."[107]

It took Railton over a week to write directly to Du Cane, on 15th April, taking care to express himself diplomatically:

> Many thanks indeed for your letter of the 5th. I was most interested in all you had to say and particularly in the photograph of the model in action. From the way it is running, it looks as if the resistance figure should be very good, though I bet the pontoons cause a bit of fuss before they come unstuck. As you say, it is a pity about the increased beam forward, at least from the aerodynamic viewpoint. The one surprising feature of the model which John is bringing over is the ease with which the forepart unsticks, though this is not true of the stern which is rather sticky.
>
> About this new model, I am delighted that you have been able to arrange with Gawn to test it. Thinking it might be the civil thing to do, I have written him direct, as per the enclosed copy, briefly acquainting him with the genesis of the scheme, and making it clear that not very elaborate data is required at this stage.
>
> I shall be interested to hear your impressions when you have seen it. As you will appreciate, it constitutes a serious attempt to provide reasonable sea-worthiness, immunity from chine-tripping, and also a not too difficult structural problem. The peculiar shallow keel in the centre of the planing surfaces is peculiar to all Van Patten's boats and he sets great store by it. He claims it provides better directional stability than a considerable degree of dead rise at the step, and at less expense in resistance.
>
> As regards the aerodynamic characteristics, I think it will prove (after a little trial and error, perhaps) to be just acceptable, but only just. What I am least confident about is its hydrodynamic drag at top speed. I'm sure this will be less favourable than the *Bluebird*, but I am naturally hoping it will not be prohibitive.
>
> Re. Fins. I would far sooner not fool with air fins if they are avoidable, as I hope they may be. Anyway, we can't tell anything until we have some yaw tests made in the wind tunnel. However, it remains a fact that there is a limit to the speed at which aerodynamic directional stability can be safely controlled by a water fin. There comes a point

> where the couple caused by the lateral air pressure on the body resisted by the water pressure on the fin (about three feet below it) might capsize the boat. Whether we are anywhere near this condition at 250 mph and with reasonable angles of yaw remains to be seen.
>
> I don't quite know what Gawn's reaction will be to testing the model without its streamlined superstructure. As I have said in my letter to him, all we really want to know at present is whether the resistance is good or bad, and I think he should be able to tell this from the model as is. If you agree, please urge this on him. I hope you will notify me of his results directly they are available, particularly a direct comparison with the *Bluebird* which is the only real yardstick we have.[108]

In his separate letter to Richard Gawn at the Admiralty Experiment Works, it is obvious that Railton did not want to risk any confusion in what was required, in an effort to speed up the model test stages.

> I understand from Peter Du Cane that you have kindly consented to conduct some trials of a model which I have been having made over here. In order to avoid any misunderstanding, and to save Du Cane the trouble of relaying it, I am taking the liberty of writing you some information concerning this model which John Cobb will be bringing back with him shortly.
>
> Since I saw you last year my mind has been rather active on the subject which we then discussed. As you know, the big problem is to evolve a general shape of hull that shall have an acceptable behaviour aerodynamically. In order to achieve this I think it is almost essential to keep the forward sections as narrow-gutted as possible, and this in turn limits very stringently the width of the forward planing surfaces.
>
> The logical answer to narrow planing surfaces seemed to be to have more of them; i.e. to revert to the multiple step hydroplane.
>
> Now it so happens that there has recently been built over here some remarkably successful 3-plane boats, and it also happens that their designer, Douglas Van Patten, is an old acquaintance of mine. It therefore seemed a good idea to find out from him what 'tricks' (if any) there were in getting a low resistance from a multi-plane boat, and to see if it was possible to adapt these tricks to our special requirements.
>
> I have found Van Patten very cooperative, and together we have

TENSIONS RISE

cooked up a set of lines which are now incorporated in the model referred to. Being familiar with the aerodynamic requirements *I was myself responsible for the general shape of the hull and for the general disposition of the planing surfaces* [author's emphasis], while Van Patten has decided the size and contours of the planing surfaces themselves.

Since one of my chief concerns with this type of hull has, all along, been its ability to 'unstick' with the thrust available, and since we had the use of a small tank with a carriage speed of 10 mph, I have taken the opportunity to check the performance over the 'hump'. The scale of the model is 1/16, and we were therefore able to run up to a full-scale speed of 40 mph. On this basis the maximum full-scale resistance proved to be around 3,100 lbs at 25 mph, with a total weight of 6,500 lbs. This gives us a good margin under our anticipated available thrust of 5,000 lbs, and I think warrants further trials to ascertain the high-speed resistance.

The model as you will receive it is ballasted to give the expected optimum C.G. position and a full-scale weight of 6,500 lbs. The thrust line is horizontal and is located at the height of the towing eye fixed in the stem. It was towed by a single line up to 10 mph, and showed no signs of yawing up to this speed.

The model has neither fin nor rudder, although both will of course be required on the boat. I believe that you agree that it is better to test without these appendages, and to make arbitrary allowances for them afterwards.

I am sorry to say that the model is incomplete in one respect. In order to save dollars we did not attempt to simulate the superstructure, and it represents, in effect, the lower half of the boat only. This was perfectly all right for our slow-speed tests here, where the windage was negligible, but I realise that the situation will be different with you. If you should decide that it is a waste of time to tow it at high speed in its present incomplete form, I think it should be possible to add a shell to give it the true shape above the waterline.

On the other hand, if you think that, by testing it as it is, you could place it in one of two categories, namely 'hopeless' or 'promising', then I think it would be worth doing for the following reasons. Its underwater surfaces in their present form differ widely from those on the model which was found to be acceptable aerodynamically. I have naturally borne the wind-tunnel findings very carefully in mind, and I think that this new form will be satisfactory from this point of view, but cannot be sure until it is tried in the tunnel. If therefore you can find out enough from the model in its present form to say that it is 'promising', the next logical step will be to try the completed form in the wind tunnel, to make what further modifications appear necessary, and finally to come to you again with a completed model for a serious forecast of performance. Alternatively, if it appears 'hopeless' in your preliminary test, we should at least be wasting no more time on it.

I shall anxiously await your verdict, and if it would not be too much trouble I should be grateful for any quantitative data you could furnish by direct comparison with the *Blue Bird* hull. I could then make a rough guess at the allowance to be made for different air-drags.

Hoping to have the opportunity of seeing you early in July.[109]

Gawn responded directly to Railton, only to confirm that he was pleased to test the model as had been instructed, but interestingly adding: "I quite see what you are getting at with the fins, both aero' and hydro', but agree that at the scale of this first model, it is best to work without them."[110]

On the other hand, Du Cane's response to Railton seems to indicate that he had not grasped the principles of the fins or the multiple steps:

Thank you for your interesting letter enclosing one written direct to Gawn, which will be very helpful.

I am sure Gawn will be able to manage without the fairing for the time being, but if desirable we will do the necessary.

To refer to the air fins for one moment, while I can perhaps see your point, I still would not have thought that fins low down masked laterally by the hull or fairing would have been of much value. Considering any one fin, if you consider the boat yawing either to left or right off its course you will see that the fin in each case is masked by the lateral surface of the hull adjacent to it. Surely a fin, if required for this purpose, should be placed on top of the hull on the centre line disposed abaft the lateral centre of pressure.

However, these are minor details and I look forward to many interesting arguments when you come over again.[111]

TENSIONS RISE

There would be no need for any discussions when Vosper received the model and Du Cane realised what he had been missing. Cobb delivered the model on 26th April and Du Cane wrote to Railton that same day:

> John handed me over the model today, which is certainly a beautiful piece of work. I now see what you were getting at over the question of the fins, which was not entirely clear to me from the drawing.
>
> Gawn has agreed to test the model as soon as possible, but regrets that it may not be until the middle of May before he can get any results. Apparently he has a very full list of visitors between now and then.
>
> Your model certainly appears likely to get over the hump easier. But from the look of the planing surfaces would appear to be likely to be more resistful as of course we are planing on three very small surfaces at the moment. In any case it will be very interesting to see and I now tumble to what you are driving at.[112]

Writing from his hotel in New York, Cobb clearly indicates his approval for bringing Douglas Van Patten on board. More telling is the clear message that after just one meeting, Cobb found Van Patten more accommodating and open than some others involved in the design process.

Six days earlier, on 20th April, Railton had written to Cobb with his in-depth assessment of the situation. He did not pull any punches:

> I enclose a copy of a letter I have written to Peter. I think you might be wise keeping a fairly close eye on what happens down there between now and when I arrive. Peter has sent me a photograph of his latest model, however good the high-speed resistance figures may be, I am perfectly convinced that it would have characteristics which would be dubious from the safety point of view.
>
> What I am getting at is that, unless this new model of ours shows a prohibitively high resistance at speed or is quite unexpectedly bad in the wind tunnel, I am certain that it will be a great deal more practical than anything I have seen from Vosper. Needless to say I have so far scrupulously avoided saying anything of this sort to Peter.
>
> When and if, therefore, Gawn decides that it is 'promising', there is no doubt in my mind that you ought immediately to have a wind tunnel model made, regardless of the results on the Vosper model (unless of

CRUSADER

> **HOTEL GOVERNOR CLINTON**
> IN THE PENN ZONE
> 7TH AVENUE AT 31ST STREET
> OPPOSITE PENNSYLVANIA R. R. STATION
> BALTIMORE & OHIO MOTOR COACHES STOP AT DOOR
> A MEYER HOTEL
> NEW YORK 1, N.Y.
>
> 18. 4. 50.
>
> Dear Reid
>
> Van Patten did not turn up until Monday after all and we had a very pleasant twelve hours together, we got on first class and I like him. I made him the alternative offers and he seemed pretty definitely to favour acceptance of $500 which I gave him, and he said he would come over to England at his own expense if wanted, provided I took care of him in England and paid his return fare. He certainly is very keen. I enclose $/65 for Model to keep accounts straight. I hope you have had some success with the Cunard Co. No doubt you will send Peter drawings for Wind Tunnel Model with all instructions

One of Railton's many letters urging Cobb to "keep after" Vosper and Peter Du Cane. It is evident that by now he was not only frustrated but also concerned that Vosper was drifting from the project's aim. *Railton/Cobb archives* course they have something good that I know nothing about). I'm afraid this may prove an awkward situation for you, particularly as Vosper would normally make the model, and if (as may well be) the high-speed resistance of Peter's model is appreciably less than ours, because Gawn will then naturally be in favour of it, since he is not concerned with aspects of safety, ease of construction etc, and would probably take Peter's opinion on these points.

When the trial is made, will you see that the figures are sent to me immediately, so that I can form an independent opinion of the situation?[113]

Cobb wrote to Railton on 27th April to confirm he had arrived home safely with the Van Patten model and had delivered it to Du Cane the previous day, adding: "His [Du Cane's] own model now looks almost identical with the original wind tunnel model as okayed by Fairey's, except that he has now got two long narrow steps and has shaved off the large blister which he had on it when we went away. As you know I am no judge of these things, but compared to your model his wheel base is very short. I will keep after him and see that you get your figures without delay, but I shall be very happy when you are over here to take command."[114]

Cobb's next letter to Railton was a brief one on 11th May 1950, in response to communications from Du Cane: "Nothing much to report to you. I notice from copies of correspondence that Peter queried the C/G [centre of gravity] of your model. I had already informed him that it was correct, as Van Patten told me in New York that you particularly did not want it altered."[115]

Railton replied on 26th May, bemoaning the fact that he had still yet to receive any data from Haslar regarding the Van Patten model, although he was convinced they had run it by that time. He was also convinced that any figures recorded would have been "fairly favourable".[116]

Whatever the schedule at Haslar, it was not until 23rd June, four weeks later, that Cobb cabled Railton with the message: "Short talk Peter — Indicates evidence very satisfactory model — Unable elaborate now."

A break appears in the written archive at this point, as Railton made one of his visits to Britain. Even though Cobb was looking forward to his friend "taking command", it appears that things instantly slowed down again after Railton's departure, as Cobb informed Railton in a letter dated 3rd October: "I am afraid this letter is long overdue, but I have been waiting thinking I should hear something from Peter, but have not got very much to report in that direction, except that he is getting on with the model and assures me that the rocket is also in the course of preparation, and will not cause any delay."[117]

Railton's intense frustration is apparent from the cable Cobb received on 10th October, the day Cobb's letter arrived: "What? Still No Data? No Rockets?"[118]

It has to be assumed that this was enough to prompt Cobb to book a trans-Atlantic telephone call with Railton as there is no record of a written response. In any case, Railton followed up with Vosper fairly quickly, as confirmed in a letter to Cobb on 17th November: "I enclose a copy of a letter to Peter which explains itself. There is no doubt that the boys over here [in the United States] have really got something. I believe these 'propriders' are going to be in the 200–250 M.P.H. range during the next few years, though (like ourselves) they will have to change their aerodynamic shape quite a bit."[119]

However, Railton's missive to Du Cane did not seem to spur Vosper into too much action, as the next message from Cobb to Railton on 19th December indicates: "I have not heard much from Peter of late, and on calling him this morning was informed that he had commenced his Xmas vacation. However, the rockets apparently are ready, although not entirely satisfactory, and his last letter to you explains the situation regarding the steering gear."[120]

CHAPTER 6
SLOW GOING

With the arrival of Douglas Van Patten's model at Vosper in April 1950, along with several drawings of it as well as of the original scheme devised by Reid Railton with Thomson & Taylor, the models being tested by the boatbuilding company finally started to take on a more unified appearance.

At the start of 1951, model 'CJK' looked quite similar to *Crusader*'s final shape. The rear-mounted sponsons, or floats, were further back, separated from the hull by raked support arms, and the model included features that can be seen in Railton's drawings and in his model, in the drawings of Van Patten, and in the hybrid Van Patten/Railton model. Especially notable was the double-chine forward planing shoe, with its two planing surfaces of differing heights, the hull first rising on the wider outer chine, then planing at speed on the narrower inner chine.

At about this time, the model's performance began to improve considerably, and the letters travelling between Reid Railton, John Cobb and Peter Du Cane became less strained. In a brief hand-written note to Cobb from The Whittier Hotel in Detroit, Railton seemed almost relieved after offering New Year's greetings:

> Gratifying to hear all the elements are coming together at last with the model. Quite why Peter resisted the separation of the floats from the hull, and the stepped underside (more accurately double chine element) of the bow, is still beyond me, but obviously the figures didn't lie, not that I have seen any as of yet! If you are due to call, perhaps you could

rattle his cage in that direction?

I do however have to agree with Peter with the basic aero' shape, but I had had some thoughts on it when we moved your perch, from the bow to mid-ships, as to how we get the air into the engine, but perhaps he has something up his sleeve for this?

If we can now just persuade him that the sponson arms need to be straight and form part of the engine frames, we can finally move on.[121]

There is no record of any other letters until 3rd February 1951, when Railton wrote separately to both Du Cane and Cobb. In the letter to Cobb, he felt the need to elaborate on his previous note:

I duly heard from Peter confirming the figures you sent me, but giving no other details to speak of. However, the bare figures of speed, thrust and weight have given me plenty to think about, and I enclose a copy of a letter I am writing him today which will indicate my thoughts.

I think the experiment shows that the major problems of stability have been pretty well solved, but the hydrodynamic resistance (i.e. the drag in the water) is still far too high to warrant us saying that we have found the whole answer.

There is no doubt in my mind that we have still 'got the side brake on', and I should personally feel very dissatisfied if we relaxed our efforts at the present stage. If we knew what the trouble was and decided that it was unavoidable, it would be another matter. Also if you had personal reasons for making an attempt before a particular date, it would be a different matter. But from a purely technical point of view there is still a bug in the thing which ought to be cleared up before we think of scaling it up to a full-size boat.

I don't mean by all this to depreciate what we have done so far. To get that model to run free at 97 M.P.H. with apparently perfect behaviour is really quite a step forward in the art, and I give Peter full marks for his share in it. It is this very striking success which makes me all the more keen to <u>complete</u> the solution.

In the letter to Peter I have asked him to send me some further information. He is very apt to say he hasn't got it or can't get it. Will you needle him about this? I might be able to deduce some clue to the trouble, and have a good opportunity hereof studying the thing at leisure.[122]

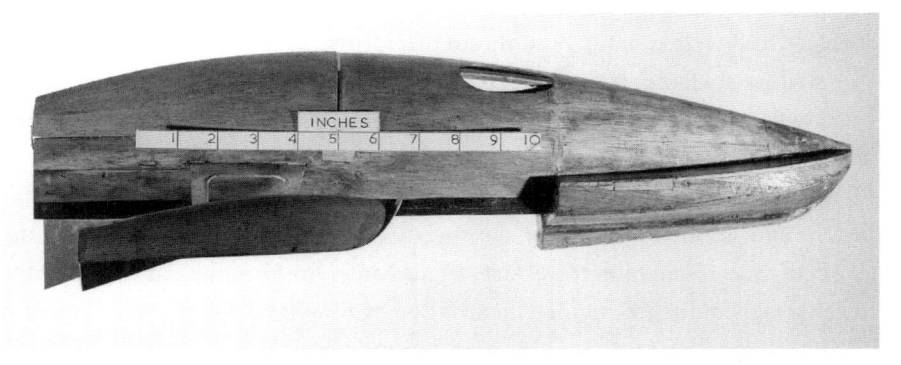

Model CDQ (a development of model CJK) shows its hybrid nature derived from the Railton, Van Patten and Vosper models. Railton's double-chine system is clear at the bow and the rear sponsons have been raised from the CJK position. *Courtesy of BAE Systems (formerly Vosper Thornycroft)*

Cobb's reply to this contained one key detail, that he was under no pressure to attempt the record before any particular date, "not this year or next my friend".[123] As for Railton's letter to Du Cane, it was indeed a searching one. Despite apologising that he might come across as terse in keeping the letter short, Railton wrote at length over seven pages. Not only did he request from Du Cane a mass of data and outline how it should be presented, but, for the first time, he began to question the key elements of the planing surfaces.

The building of a boat for the water speed record was, and still is, fraught with scientific problems and engineering compromises. The craft needs to house the engine, its fuel and the pilot in a shell strong enough to withstand the forces it will encounter at high speed and be safe enough to protect the pilot, while still being capable of floating. It needs maximum power to go as fast as possible but also flexibility of power delivery in order to make the transition from rest 'in' the water to speed 'on' the water, to get 'unstuck' — or, as Railton put it, 'over the hump' — and onto the planing surfaces, thus reducing drag and allowing speed to increase.

The term 'over the hump' is a well-known one with hydroplanes and refers to the typical shape of a planing craft's graph of resistance against speed. During acceleration from rest, drag initially rises slowly but then starts to increase steeply and reaches its maximum as the craft tries to lift itself onto the plane, then falls off as planing occurs. The only way to achieve this is to angle the planing surfaces so that the boat rises as it gathers speed, leaving less of the actual hull in the water. As the craft progressively rises up the angle of its planing surfaces, the wetted areas become smaller. If the angles, sizes and profiles of the planing surfaces are incorrect, support will finally break down and the boat will drop slightly, then resume the climb — a phenomenon described as porpoising in pitch. In some circumstances the hull could seemingly rock forward and backward rather than rise and fall vertically. Railton could see this becoming a serious problem at the speeds he and Cobb were contemplating, especially after watching the efforts of Sir Malcolm Campbell at considerably lower speeds in the jet-engined incarnation of *Blue Bird K4*.

The planing surfaces on the models tested by Vosper were flat, as was necessarily the case when working at small scale. As early as possible in development, Railton wanted to address the problems of a single angle of attack as well as 'deadrise' — the 'V' angle a hull presents to the water (as seen from the front) running upwards, usually from a central keel to the chine. In his 3rd February missive to Du Cane he added a postscript that stated:

> Since writing the attached letter I have determined to put before you one other consideration which has been fermenting in my mind for some time. It concerns the angle of incidence of the planing surfaces. Such conclusions as have been published on the optimum angle of incidence are rather conflicting and show considerable differences of opinion. These various disagreements seem at least to show the optimum angle may vary with the conditions, i.e. with the unit-loading and <u>with the speed.</u>
>
> Now at 100 mph (1/6th scale) we are operating at speeds far in

excess of that at which any experiments have previously been made, and it may be that our present planing angle is far greater than it should be at this speed. I am aware that, if we changed the angle of our present surfaces to, say, 2 degrees, we might not get the lift necessary to get over the hump and to operate satisfactorily at low speeds. However, if, instead of making the longitudinal contour of the planing surfaces straight, we made it follow the arc of a circle (as in the attached sketch), we could get an effective planing angle of 2 degrees at the heel, where the boat rides at high speed, and still retain our present mean angle to give the lift required from lower speeds.[124]

At a stroke, Railton had removed one of the hurdles facing the project.

Meanwhile, Railton's in-depth questionnaire and request for data seems to have acted as another catalyst. Du Cane's reply of 13/14th February was copied to Cobb:

Thank you for your letter of February 3rd, which certainly gives me plenty to think about. Frankly I cannot attempt to answer fully except over a course of time as the data you require arises in general terms, however, I can assure you that our initial success has not blinded me to the fact that it need not of necessity be the optimum from the performance point of view.

Can I perhaps re-assure you by saying that I have a programme of experiments in hand to test various step arrangements, hull forms, rudder arrangements and fins? Your suggestion as regards the angle of incidence is a good one and will also be tested.

Unfortunately, although it may sound stupid, I have no drawing of the model we ran in the initial test, but it is broadly speaking the same as what might be termed the 'long' model which was here when you arrived last year and is quite likely the worst for resistance but very good for behaviour.

We shall be testing the later one in due course. This may prove better, but an indication from 1/16th scale run did not point to this. However, we shall leave nothing to chance which can be eliminated. Shall also test the wider step as suggested by you last summer.

Your note about the angles of incidence is almost certainly correct, but you will be the first to realise that we have had to arrange a compromise here in the course of the 1/16th scale running, and frankly these incidences have so far proved to be the best all round. Shifts of C.G. will be tested.

I will report if we obtain favourable results from your suggestion as soon as it is possible to try it.

You will doubtless realise these 1/6th scale model tests are rather difficult to stage, owing to weather and other factors, so that I may not be able to give you the answer for a month or six weeks.

I am not convinced that the Apel form is really so much better, if at all; a tremendous lot depends on the appendage drag in each case and this is where Gawn is busy.

It must be rather maddening for you to be missing this stage but don't worry we are plodding away and your visit in May will be plenty of time to assess all data obtained.[125]

The most interesting part of Du Cane's letter is the short line stating: "Shall also test the wider step as suggested by you last summer." When one considers that Railton's letter was dated 3rd February, it is plain to see how Railton and Cobb were feeling frustrated, as the broad step suggested by Railton had been tested in Michigan nine months previously, and the drawings and data for it sent to Vosper on 24th June, five months before Du Cane mentioning that he was about to test it. It is also difficult to reconcile how Vosper test data existed "only in general terms".

Within days, Cobb received a cable from Railton: "PDC about to test multistep? Five months late!"[126]

According to Cobb's office diary, he visited Vosper soon afterwards. What transpired can only be guessed at, but it spurred Du Cane into revisiting Railton's letter and responding again with a bit more thought. Copies of this letter survive in three separate archives and, strangely, with two different dates, 15th and 22nd February:

I still feel in view of your very long and painstaking letter I have perhaps not dealt at sufficient length with some aspects of it.

The figures I gave you for the 1/16th scale model, where an equivalent thrust of 4,000 lbs in a 6,700 lbs displacement model passed the 'hump' quite satisfactorily, should be reassuring.

As regards the question of the form of the curve of hydrodynamic

SLOW GOING

The Haslar test-tank facility. Above each of the long water tanks, a suspension carriage was mounted on full-length rails. Test models would be hung from these carriages, attached in such a way as to replicate the proposed thrust line, and the model would then be towed forward. The facility was used to test in scale all manner of craft, from rowing boats to oil tankers. Courtesy of BAE Systems (formerly Vosper Thornycroft)

resistance, it is true that there is some evidence from some *Bluebird* data that the H.R. [hydrodynamic resistance] remained constant throughout a certain range. This, I submit is misleading because there was a 'virtual' reduction in displacement due to aerodynamic lift.

For reasons of safety we have all along decided to eliminate the element of aerodynamic lift and as far as is known at present neutral lift characteristics have been achieved up to substantial angles of attack.

Without lift a form of this general type will rely upon its [wetted] area-reducing qualities for low hydrodynamic resistance at speed.

Our form as developed seems reasonably good in this respect subject to refinements such as angle of incidences as suggested by you and adjustments to the C. of G.

BUT it will not have the assistance of aerodynamic lift, so as I see it the H.R. must climb with speed.

The appendage drag is excessive because I hope to achieve satisfactory control by one rudder forward combined with two smaller fins aft in full scale.

Aerodynamic drag can undoubtedly be reduced but there are some points here for settlement. The rest would have to rest mainly with relieving the planing surfaces of load by incorporating lift, a process I am reluctant to indulge in here as the form is a new one and in any case the thrust is available if necessary.

The modification to angle of incidence as suggested by you has been incorporated on the 1/16th scale model and run well. This was done just to ascertain no serious alteration in behaviour was involved, as it would have been a pity to risk the large model.

I made a slight alteration to the profile, as instead of a pure circular form I incorporated an element of the venturi or one side of a nozzle, which in theory facilitates the acceleration of the particles of water from forward to aft with minimum resistance. We can, of course, very readily try the circular form if necessary.[127]

Of course, Railton had stated from the very outset that the hull shape should be "aerodynamically neutral, creating neither downforce nor lift", the same design philosophy that he had followed on Cobb's land speed record car.

Perhaps Railton's biggest problem in his communications with Peter Du Cane was his need, as he put it to John Cobb, "to tread lightly, as if on eggshells, he does rather take offence to anything he hasn't considered". But even when sent drawings and direct requests to try certain modifications, as evidenced by the time it took to decide to test the wider forward step, progress could suddenly become very slow, and data scarce. Furthermore, it appears that Cobb also felt Vosper was making decisions without necessarily listening to those who had commissioned the work. In a letter dated 26th February, Cobb wrote to Railton stating:

Peter is difficult and your letters are just the right medicine for him. I am standing firm for 250 mph with 5,000 lbs thrust and he is accusing me of 'raising my sights', which of course has an element of truth in it. The main thing is I am certain the sooner you can get here the better as I don't think he quite knows what to do next.

Stung by your taunt last year that Malcolm [Campbell] had me licked when it came to getting things done! I went to see Air Marshal Boothman, who is the new director [air] of the ministry of supply. I knew him years ago when he won the Schneider Trophy and he expressed himself flat out to help. He wrote to Halford to this effect and I think it has done the trick.[128]

Railton's response, dated 1st March, confirms that both men were thinking along the same lines:

Thanks for your letter of the 26th. As a result of Peter's letters I had come to the same conclusion as yourself namely that the sooner I get over there the better. Accordingly, I have reshuffled my arrangements, and am now planning to catch the *Queen Mary* on March 30th.

Good news on Boothman! Frank had the engine in place when we did the first scheme, so a Ministry letter must have tipped the board over the edge to agree. I have a GA [general arrangement] drawing already that he sent me, so we already know it will fit, depending on where Peter has you sitting now![129]

Cobb was delighted to receive confirmation of Railton's next visit to Britain. In his brief reply to his friend on 9th March, it is evident he had been having some quite difficult conversations with Du Cane and felt he was being 'guided' away from some of Railton's reasoning: "I am sure your presence is the only way to get things moving as Peter will not argue [with you] that the drag is excessive."[130]

Railton penned a quick hand-written note in response, dated 15th March and sent to Cobb's home: "I had hopes of leaving them to it, but it appears not. The project manager needs managing! Why Peter is arguing is beyond me, I have sent him the figures. I also stressed that the reliance on the Froude figures [in naval architecture the Froude number is a significant figure used to determine the resistance of a partially submerged object moving in water] is a dead end, as these are for displacement vessels, which ours is most certainly not!! Now I hear [Maurice] Hooper [aerodynamicist at Fairey] is altering the planing surfaces and Gawn has moved the c/g, against firm instruction not to. I fear the *Queen Mary* won't get me there fast enough!"[131]

Railton may have been misled here, as Vosper-trained marine architect Lorne Campbell subsequently pointed out: "The Froude relationship certainly

CRUSADER

The *Crusader* team pays a visit to Haslar. In the front row are (from left) Peter Du Cane, John Cobb, Sir Charles S. Lillicrap (Director of Naval Construction), Reid Railton and George Eyston. *Courtesy of BAE Systems (formerly Vosper Thornycroft)*

came about because of work on displacement craft, but it has been found to work for planing craft, too — shown mathematically and not just by practical observation! The ⅙th model at 97.5 mph scales to a speed of [approximately] 240 mph for the full-size craft — and this is by the Froude relationship. It is used throughout the marine industry and in tank testing of both displacement and planing craft."[132]

Upon Railton's arrival in Britain, it is difficult to ascertain who spoke to whom and about what, and the Vosper archive shows that there seemed to be an air of confusion. For example, there was a point at which the 1/16-scale model had to be rebuilt, probably after a test on 25th April, although there are conflicting references to this work having been done on 21st February, 9th April or 10th April. Whichever, it was decided that during the rebuild the aft-mounted rudder would be moved to a forward position, behind the leading planing surface.

With Railton 'on site', there seems to have been an increase in activity, including more model tests. On 20th June, three days before Railton's return to the United States, Du Cane laid on an "interesting rig test" and a viewing of the most recent test films. The next day, Railton went to Thomson & Taylor at Brooklands and met Reg Beauchamp, whose recollections give some clues to Railton's opinions at that time:

> Reid called in to see me and conversation gravitated toward Mr. Cobb's high-speed boat. I'd had nothing to do with it since doing the stress calcs for the frame, and drawing up the model. With Reid, nothing was dressed up, he said what he felt needed saying, which could appear rude, but it wasn't meant to be, it was just efficient!
>
> He told me that Vosper and Gawn felt they had enough data to start construction of the full-size craft, but he felt they hadn't considered some crucial elements, and told me he had told them so. He said something like, "It's very close Reg, but they haven't tested the support arms for the floats, nor how they will be incorporated, as they don't seem interested in the armature you drew up for me. Nor have they considered the full forces on the full-size planing surfaces, which are still as flat as pancakes, and the rudder is simply wrong.[133]

Vosper for its part also recognised that Railton had a point. In his interim report some eight months earlier, Railton had written that the support arms would need to be straight, and without any rake toward the aft of the craft, as this would cause lift. Including a sketch of his original armature design, he had pointed out that the engine-mounting hoops, which combined the supports, would be the best solution, and if they dropped from the hull to the sponsons at the correct angle, which he had tested in the Michigan water tank, they would not interfere with the transition stage when accelerating. Surprisingly, it was only now, at this late stage, that Vosper decided to test the arms and their positions. Railton had obviously repeated his earlier concerns regarding the flat planing surfaces. Originally, Du Cane had said that there was no point in testing this until the model tests moved to the $\frac{1}{6}$-scale model, but this had not been done. In his first letter to Cobb on his return to the United States, probably written onboard ship in the last few days of June, Railton wrote:

> While I am a little annoyed that Peter has only just caught on to the engine/float mounting requirement, hence the reason the Ghost may not fit in the shape currently being tested (although we weren't in a position to confirm Halford had pried a Ghost free), I am lost for words as to

why there is absolutely no data for the Vee-ing of the planing surfaces. I told Peter he has greatly underestimated the forces these surfaces will encounter, especially the forward one, but his answer was that the planes would be narrower on the full-size craft, and were crucial for getting over the hump. So much for being at the coal face!!!'[134]

Cobb did not write to Railton again until 25th September, so it seems likely that in the interim the two men discussed matters on the telephone. Cobb stated:

I am afraid I have not written to you for a long time, but I have been hoping, vainly as it turns out, to be able to have a little more to tell you.

I received the cutting, which I am sending on to Peter. Possibly it may stimulate him a bit, which would be a good thing, as he certainly has not made any great progress since you left. He is going ahead with the design of the hull but is still playing about with the sponsons. At your suggestion I believe he has made them now with vee bottoms and, somewhat unexpectedly I think, has run up against some snags. And tells me that he can only cure this by widening the front step, but he has not satisfied himself to date. I spent the day with Peter, and he does not seem to have much idea of how to arrange the steering gear, and suggests co-opting Lockheed's with a view to making it power operated.

Altogether I do not feel too happy about the progress which has been made. As Vosper are still in the red, I suspect the banks have a say where private orders are concerned in case they are not paid for, anyhow at Peter's urgent request I have had to pay them a deposit of £5,000!'[135]

The small amount of surviving written communication during Railton's visit to Britain includes a last-minute letter to him from Du Cane, written on 22nd June, enclosing a blueprint. In explanation, Du Cane stated: "As you know quite well it by no means represents the final shape. Any bright ideas you might have, especially as regards detachability, will be welcome."[136] As Railton was sailing to New York the next day on the SS *Wedia*, Du Cane addressed the letter to him care of Cunard White Star in Liverpool.

This demonstrates that only now was Vosper looking into producing a full set of construction drawings and determining how the vessel would be transported — even though Railton had assessed the problems involved in

Richard Gawn headed the test-tank facility at Haslar. *Courtesy of BAE Systems (formerly Vosper Thornycroft)*

both construction and transportation a full 16 months previously. A brief hand-written note to Cobb again shows how frustrating Railton found this:

> Received the enclosed from Peter. The drawing shows development swinging to our way of thinking, but Peter's resistance is still evident.
>
> You recall I asked for the largest permissible width for road transportation in Feb 1950, and it has now occurred to Peter to ask how the engine, and sponsons, might be mounted, and the latter detached for transport, yet he confirmed to me when I was over he had received the drawings for all of these in March (1950)!! I also see in the drawings that he has sent, that although the shape is now closer to ours, there is no inclusion for the alloy armature. It would seem that as he slowly accepts our rather unconventional and futuristic design, he is still clinging to convention.[137]

Railton's next task was to reply to Du Cane, and remain diplomatic, while again trying to steer the project in what he believed was the right direction. He sent the following letter from The Whittier Hotel in Detroit on 2nd July:

As promised I send you herewith my suggestions for the attachment of the sponsons. It is an all-steel construction. The same thing could be done in Dural, but the members would have to be designed by a competent aircraft man.

The parts are designed for a maximum shock load of 10g, and for a material having a yield point of not less than 100,000 lbs/sq in. The construction is such that the members could be made quite easily by a firm like John Thompson Ltd. In the old days I should have expected to get them f. o. c., and still should if my old friend Mr Aston is still with them.

You will note that they are readily detachable by the removal of four bolts. Moreover, the attachment to the hull is at four points only and well below the centre-line of the engine. No 'strong member' is required above these points, which should ease your construction problem somewhat.

There is obviously no room for this suggested construction at stations in way of the main body of the engine. If you use it, therefore, one member would have to come in front of, and the other behind, the main mass of the engine, as in:

Fig 1. The weight of the four arms as shown is around 300 lbs. total. This is on the heavy side, but is, I think, justified by the simple and light construction which it might enable you to use for the hull and sponsons. The arms could undoubtedly be made lighter in Dural, using riveted sections, but it would certainly be a lot more expensive. Also steel is far easier to repair if damaged.

Fig 2. is intended to be a sketch showing the general arrangement of one of the four arms. I think it is self-explanatory. They could be sheathed with light aluminium fairings for appearance and reduced windage, though one would not bother with this at first.

In the interests of cost it would be most desirable to keep the shape of all four arms the same, so that only one forming block would be required. I think this should be possible by merely varying the length of the forward and after arms.

Figs 3, 4 & 5. are detailed sketches showing the proposed construction. The arms themselves are two channel-members welded edge to edge and forming a rectangular tubular section. This welding need not be continuous though there should be as much as is consistent with freedom from distortion.

I think I have shown enough of the scheme to enable you to try it out on paper. If you decide to go further with it, I suggest you send me the final drawings of the steel parts when they are ready. I can then check them and take that bit of responsibility off your shoulders.

One final word about vee-ing the planing surfaces of the sponsons as we discussed. Personally I would far rather risk initial troubles and alterations in getting over the hump than face the possible results of shock loads on a dead flat surface. In other words, I would take a chance and vee them, even if model experiments were inconclusive [author's emphasis].

Let me know how things go, and use my California address until further notice.[138]

This letter confirms that Vosper had yet to incorporate two significant elements of Railton's design work into the models. First, the company was still using the raked and angled sponson supports that Railton felt were causing the ⅙-scale model to become unstable, by creating an aerodynamic effect that lifted the stern of the model out of the water. Second, the need for some 'vee' or dihedral on the planing surfaces had not been instigated. Railton's letter also dispels later claims that Railton had sent a mere 'sketch' of an idea.

Railton later related to Reg Beauchamp in a visit to him that he had drawn Du Cane's attention to his thoughts on the matter: "Reid felt very frustrated, but slowly but surely Vosper were coming around to what we had already worked out, but they appeared resistant to change. At first they didn't like the reverse three-point layout, arguing that the wash from the forward shoe would affect the two rear shoes, but came around to it, and it was the same with everything else. Reid knew the various components worked, but getting Vosper to test them took an age."[139]

Railton's presence, however, did have some impact on the model tests, as a new model, 'CJQ', had been made. This, in effect, *was* the Railton/Van Patten hybrid design, minus a full '1½' forward step and sponson fins. In profile, its shape was all but the final one, including the original forward cockpit concept that Railton had previously dropped.

Du Cane's next letter to Railton gave some response to Railton's concerns about engine/sponson mounting and planing surface vee-ing: "I have been in touch with the Chief Engineer at Supermarine, who has given excellent co-operation, including design advice, and has shown us how to solve a very similar problem in one of their latest fighters, the 'Swift'. In the case of the tail unit here, the maximum stressing is very similar and the spar is also detachable. They use a built-up ring structure composed of 28 tons/sq in. alloy (DTD 610). I see no likelihood of real difficulty over the vee-ing of the rear sponsons and am working on these lines."[140]

Again this seemed to irk Railton, who fired off one of his short, terse, hand-written notes to Cobb: "Peter has written again, and seems to have forgotten all that I have sent to him. We had the engine and sponson mountings in hand over a year ago, and it was sent to Peter, who has now come up with exactly the same thing, even the material Ewan [Corlett] had recommended, though he has incorrectly stated it as 28 tons/sq in alloy, when Ewan gave the spec' as 32 tons per square inch. Why quote under the correct figure? Peter also now claims that vee-ing the plane surfaces is not a problem, so inch by inch, he is using our data and drawings, which begs the question why the resistance to our scheme while doggedly and continuously testing old hat!"[141]

Before Cobb could reply, Railton received another letter from Du Cane (dated 30th August) that mainly discussed the 'prop-rider' *Slo-mo-shun*, but included this information: "Not much worth reporting about John's boat at the moment. We are still awaiting a final confirmation that the engine will in fact be available, and are carrying on with some structural test pieces in the meantime. We are of course, also arranging for further tests in the tank on the ⅙th scale model to establish the magnitude of the 'hump' resistance having incorporated the vee'ed sponsons and also with two alternative shapes of cantilever. Following this there will be the necessity for one or two behaviour test runs with the 26lb. rocket installed, all of which will probably take a few weeks yet, but most material has been ordered on the assumption that engine will in fact, be allocated."[142]

So, Vosper had finally incorporated, bit by bit, nearly all the key elements of the design that Railton and Beauchamp had hammered out before the boatbuilding company had even been approached. Yet the two points that Railton had, in his own words, "sweated blood over" — the forces the hull would encounter and the means to mitigate them by means of an armature structure and vee-ing of the planing surfaces — were still meeting with resistance. Du Cane wrote the following letter to Railton on 28th September:

> The position is now gradually unfolding itself and I am writing to you because I am pretty near having to make a definite decision and would

SLOW GOING

Model CJQ illustrates the coming together of Vosper's research and Railton's thinking, although Railton had constantly warned Vosper about raking the arms for the rear sponsons. *Courtesy of BAE Systems (formerly Vosper Thornycroft)*

very much like to get your sympathy, and if possible, agreement.

Broadly speaking, to summarise, when you left I agreed with you it would be a good idea to try Vee'd sponsons associated with wider forward step as suggested by you, or narrower step incorporating some element of slip chine.

We first of all carried out runs with the ⅙th scale model in the Haslar Tank to establish resistance and behaviour at 'hump' speed. The Vee'd sponsons in association with the narrow step forward rather surprisingly proved unacceptable, the resistance was excessive and the model rolled violently to such an extent that no records could be taken. It was in other words a washout.

I then suggested trying a wide forward step, still in conjunction with the Vee'ed sponsons. This combination was successful, the 'hump' resistance being reasonable.

We then proceeded to run the model with the 31 lbs. rocket installed on the water for behaviour tests (for behaviour I thought it best to overdo it a bit). The model ran very well for about 150 yards, then took off, turned over etc. As there was some possible element of doubt as to the conditions I ran the model again on a perfect day in exactly the same conditions as regards step, sponsons, etc. etc. Once again she ran about 100 yards, and then lifted forward with consequent troubles. I am afraid I cannot escape the conclusion that there is too much lift both aerodynamically and hydrodynamically with the wide step.

I am going back now to the original arrangement of narrow forward step with the original sponsons and will incorporate a slip chine if possible.

Shall be doing some final tests when we have repaired the model, which may not be for two or three weeks.

One thing you will be rather pleased about, and that is that we have succeeded in running the model quite satisfactorily through the 'hump' and up to planing condition with a low cantilever, which greatly simplifies our structural problem.

If, therefore, we are very careful indeed in getting adequate strength in this cantilever, which of course is much easier now, I do not think there is an undue risk in using the flat sponsons, though I have appreciated your point in suggesting the Vee-ing.

Have kept John in touch and will notify you again as soon as further running has taken place.

John has sent me cutting you forwarded to him. It is all part of the day's work and I know how these things happen, but if you feel like it would like to see you over here for a few days again.

P.S. It is awfully difficult to put everything into a letter properly and can only say that have done my best to describe and if you are not happy the only thing to do is to come and talk to me.[143]

Three days after writing this letter, Du Cane sent a brief postscript: "I would like to assure you of the fact that the model is now running very satisfactorily from a directional point of view, i.e. the large fins on the sponsons are doing their work, there is no question of any trouble in this direction. Also the thrust line is faithfully recorded as between model and predicted full scale."[144]

Quite why Du Cane felt the need to ask for Railton's sympathy not to vee the planing surfaces is a question we can never answer. From the moment Railton and Cobb had first approached Vosper with concept drawings and calculations, Railton had made it clear that he knew what to expect by way of impact forces on the planing surfaces, and how they should be dissipated, and wanted them tested in model form — and still it had not happened despite reminders. Yet again, Railton found himself writing a letter to Du Cane that was necessarily forceful but still diplomatic. It was dated 3rd October and copied to Cobb.

> Many thanks for your letter of the 28th Sept. I agree that, on the basis of the data which you report, the prudent thing to do is to go ahead with the narrow, flat planing surfaces. It is still inconceivable to me that these flat surfaces will eventually prove to be the best form, in fact it is partly the derision which we shall doubtless encounter when we disclose them which sets me against them. Couldn't you get by with just a <u>little</u> vee at the forward ends?!
>
> I am a little disturbed by your statement that the model 'took off' with the wide front step. On the basis of the wind-tunnel tests it should never do this under any conditions, which makes me wonder whether we have inadvertently introduced more lift in the upper contours of the body. The Fairey model was of course quite different in its upper-works from the present one, and I think we should check on this before we go too far.
>
> It is very satisfactory that we can get by with cantilevers of a better functional shape. I shall be interested to see what method of construction you finally adopt.
>
> I heartily echo your wish that we could have a talk about things, in fact I have been looking at the possibilities of such a meeting myself. At the moment I can't think of anyone upon whom I could work off the expenses of a trip, but you can be sure that I shan't miss any chances that may come along. When do you expect to get your final proposals completed, i.e. air ducts, tanks, steering gear, fins etc? It might be a good thing to know the optimum time to make a trip, in case opportunity offered.[145]

In an article published in *Motor Boat and Yachting* in August 1952, just before *Crusader* ran on Loch Ness, Du Cane confessed that he had been impatient, with some model tests running in less-than-perfect conditions, which usually ended up with the model needing to be rebuilt over "weeks rather than days", and stating he felt that the aerodynamic lift created by the wide forward step was causing the models to take off, probably not helped by water conditions affecting the angle of attack.

No matter, Railton's latest letter crossed with Cobb's reaction to Du Cane's letter. It is quite clear that even Cobb was now concerned: "Have just seen Peter's letter to you dated 28/9/51 and it would appear that his experiments are getting him nowhere. I am wondering if you think it really necessary to vee the sponsons for safety reasons? If you do, the subsequent delay is unavoidable."[146]

On 5th October, Railton found the need to send another of his quick hand-written notes, on the back of a Hudson Cars compliments slip: "Full response to follow, but in short, I can see why Peter waited until I had I left! We have been pushing certain key elements from day one with Vosper. Indeed, it was on those grounds we went with them! Had he tested what we gave him, we would be up at Coniston now. Yet now he's resisting the Vee'ed surfaces, some two years since we presented our scheme. The same for the sponson supports and the engine frame assembly, but as of now, they remain untested. I have written back to him, but I fear there is more to follow."[147]

Railton's subsequent letter, dated 10th October, suggests that he and Cobb had then discussed the matter by telephone, as there is very little reference to the matters that had so irritated him: "Your principal question is answered by the letter I wrote Peter on the 3rd, of which I sent you a copy. The situation as he describes it doesn't make much sense to me, but it's no good my trying to interfere at this distance…" Railton goes on to state that he felt he might have to return to Britain earlier than planned to see how things were progressing, but his confidence was not high: "To make the trip really worthwhile, I reckon he [Du Cane] ought by then to have ready his schemes for (1) the sponson construction, (2) the steering gear layout and (3) the air intakes. However, I'm afraid it would only hurt his feelings to urge this on him too strongly. Actually I shall have to come (if I do) when it suits me, and trust to luck that there will be something to see."[148]

Again, this letter to Cobb crossed with Du Cane's next one to Railton, sent on 9th October. Curiously, it ignored Railton's requests for including not only the vee-ing of the planing surfaces while leaving the forward surface in its narrow form, while making no mention of the drawings that he had been sent regarding the alloy engine frame and its incorporation of the rear sponson mounts, or the forward part of it that would support the rear of the front planing surface and rudder.

> I must admit that I was rather disappointed at not being able to incorporate the vee in the aft sponsons and am quite convinced that

SLOW GOING

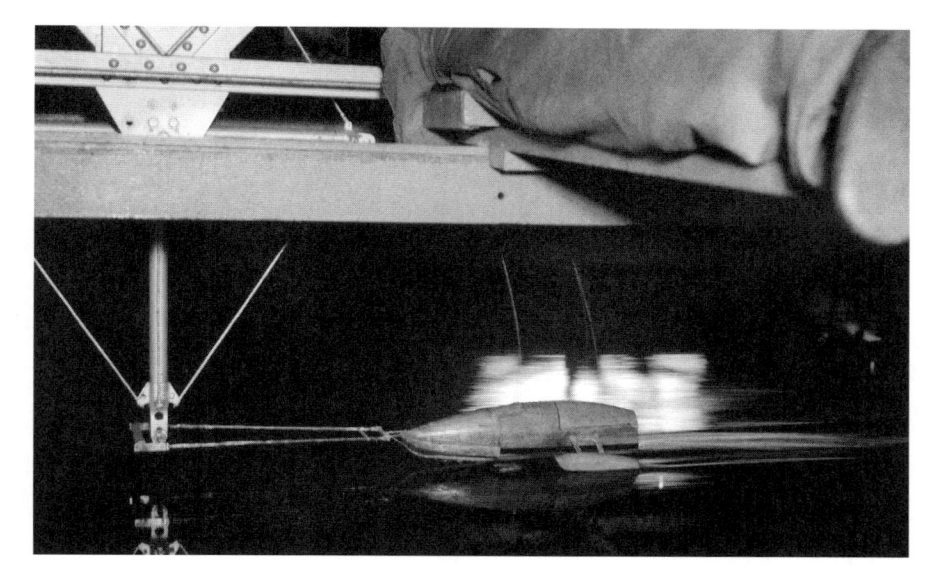

A version of model CJQ on one of the test rigs in the Haslar tank, with the towline attached at the bow along the thrust line to replicate the jet thrust. At the top a technician rides on the moving carriage as it tows the model. *Courtesy of BAE Systems (formerly Vosper Thornycroft)*

given time we could do something about this. The difficulty is that there is a rather critical balance between the requirements of getting over the 'hump' and satisfactory high-speed running, both aerodynamically and hydrodynamically.

The one we produced last winter and know, does at least appear to do its stuff and should be acceptably safe.

As regards the aerodynamic lift, I rather anticipated you would bring forward exactly the point you have, and in reply to that can only say that I did take the model to [Maurice] Hooper [aerodynamicist at Fairey] some time back, who stated he could see no reason why the aerodynamic results should be greatly different to those on the original. I am entirely happy as to the cause of our troubles with the taking off of the wide-step model, and it still might be a question of water conditions, but it really was more than I could overlook that the model did it twice. I think the probability here is, that the wide step is what might be termed 'stronger' than the narrow step, and that in the event of it hitting anything like a serious disturbance on the water the result is more violent than in the case of the narrow step.[149]

CRUSADER

These letters seem to confirm that whereas Railton felt the need to have structurally strong, vee'd planing surfaces, Vosper considered it would be the rear sponsons doing all the work, and had not tested a vee on the forward surface at all.

It also appears that Vosper intended to 'freeze' the design at the beginning of October 1951, although Railton knew nothing of this — and the archives give no reason for it. This meant that there was now no time to include and test the vee form to the planing shoes. Du Cane's letter asking for sympathy was sent far too late for Railton to be more forceful about his concerns, and it seems the matter had already been settled in Vosper's eyes.

It is strange that Du Cane, within days of stating that there were many more model tests to be done and that the sponsons had not even got to the drawing stage, went on to say that it was not possible to include any vee-ing in the planing surfaces.

As for Du Cane's comments regarding the width of the forward step and its strength, they would later take on new meaning.

Railton cabled Cobb the day he received Du Cane's latest letter: "PDC states no V on any surface. Using weaker step, and no frame. Baffled."[150]

Cobb, being the gentle giant he was, again became stuck in the middle, but ultimately it was his neck on the line. He wrote these brief comments to Railton on 15th October: "Many thanks for yours of the 10th and of course there is nothing I would like better that a visit from you. I am wracking my brains how to put it to Peter without upsetting him… I presume you know by now that he has been discussing things with Corlett?"[151]

In the August 1952 article written for *Motor Boat and Yachting*, Du Cane alluded to many of the points of contention, although his comments appear to contradict much of what he had told Railton and Cobb. For example, one paragraph reads: "Students of planing forms will perhaps observe the absence of vee or dihedral on the planing surfaces. ***It would have been better in some respects, particularly from the viewpoint of reducing peak loadings, to have some vee here*** [author's emphasis], but up to the time we had to 'freeze' the design in the interests of production." And a little later: "An advantage on the score of resistance and therefore speed can be anticipated with the flat bottoms."[152]

That would seem to explain why Du Cane was so set against the vee-ing of the planing surfaces. He had shown himself that a craft with flat surfaces would work, and would have a little less drag. He also believed that vee'd surfaces would lead to problems. As for the 'freezing' of the design, there seems to have been no real reason for this, especially as 'production' was not imminent, and the drawings had not even been signed off when he sought Railton's 'sympathy' not to include the vee form.

The article seems to show that Du Cane understood why Railton was so very keen to include a vee form, but decided that the gain in speed was preferable to the possibility of problems. Curiously, later in the article Du Cane went on to say: "In the case of the sponsons, large forces could be anticipated which would have to be transmitted via the cantilever arms to the hull, which in this region had of necessity to be as narrow as possible, while at the same time housing the turbine unit on its inside diameter. The sectional drawing will show how this problem has been tackled. In the evaluation of the loading likely to be experienced by the planing surfaces we were given much helpful advice by Mr. Peter Crewe, chief hydrodynamicist to Saunders-Roe, Ltd. Reid Railton took a strong line and very considerable interest in this point."

Again, this appears to show that Vosper considered the main baulk of the punch forces would be encountered by the sponson-mounted planing surfaces at the rear, even though Railton had continually stated that the loads at the front surface would be considerably higher. It is all very difficult to reconcile.

Cobb wrote to Railton again on 17th October, a letter that answers a question that students of this subject have often asked:

> I have contacted Peter and tactfully raised the question of your presence here, as one would expect he did not seem very enthusiastic, nevertheless I am quite certain that he would have to cooperate to the full if you decide to come. Incidentally Peter is now worrying me for a name, and the only one that sticks in my mind so far is 'CRUSADER', for no particular reason. Let me know if you get a brainwave.[153]

It now appears that Railton and Cobb had been reduced to subterfuge in order to try to persuade Du Cane to build what they had originally conceived. Railton wrote the following letter to Cobb on 20th October:

> Thank you for your two letters. I have also had letters from Peter, not noticeably enthusiastic, but apparently resigned!
> On the principle that there is no time like the present, I have settled matters amicably with Hudson's and propose to get on the

CRUSADER

Caronia October 27th, due Southampton November 3rd p.m. I plan provisionally to leave again about the end of the month.

I am writing Peter telling him when I arrive, but saying nothing of any reasons for my visit, leaving him to infer that I have some other business. I shall say that I realise that the main features of the boat must be considered settled, but that I should like to discuss some of the details such as the air-intakes, the rudder mechanism, the fuel tanks etc. I don't think there will be any trouble in that direction.

[*Author's note: Yet again, it is the hand-written postscript that is most telling*]

'Of course if Peter should contact you directly, it would be best to make no mention of the lack of inclusion of the vee-ing, engine frame, impact forces, step/rudder mount and step width etc. I'd like to see what the answers they have are, off the hoof as they say! As for Corlett, he did contact me about Peter's contact, and I simply asked him to repeat the use of the DTD 610b alloy, as he had originally suggested to us.[154]

By the time Cobb received this letter, Railton had begun his journey to Britain and within days of his arrival both men were on their way to visit Vosper. One can only guess what transpired, but it is obvious from the Vosper archive that Railton's concerns were now being addressed, as confirmed by copious pages of notes and calculations for both the shock loadings and the design of the rudder.

Of course, in view of the tragedy that was to occur on Loch Ness, this exchange of opinions and figures takes on greater importance.

Railton's 'little black book' shows that whatever figures he calculated for shock loads at the rear sponsons, he increased them by over a third (at a minimum) for the forward step — the first part of the boat to make contact with the water at speed. The notes in the Vosper archive, however, suggest that Du Cane was more concerned with shock loads on the rear sponsons if running conditions were not ideal.

As the proposed record-breaker was expected to run in near-perfect conditions, Railton used the exercise to confirm his figures for the hull structure's strength. A note from him makes reference to this:

Assuming running in a reasonable swell, if the sponsons are thrown 12' vertically (say on hitting a wave at peak design speed, e.g. 250 mph) the vertical acceleration would be 15.5 g and the load on each cantilever arm would be 46,500 lbs.

If thrown 6' upwards, this reduces to 10.9 g and 32,800 lbs. If thrown 3', 7.8 g and 23,200 lbs.

So it's my opinion that the very least we should allow for is 33,000 lbs, or 11 g for normal running, as the forward step will have quite an effect on the sponsons, because of the wake effect. Therefore, for the forward step, in normal running we should consider impact forces up to a maximum of 50,000 lbs or 22 g. Of course with a degree of vee-ing, we can reduce these at a stroke.[155]

We can only assume that Railton conveyed to Du Cane his recommendation concerning the strength required for the forward step and the notes in the Vosper archive support this assumption. Railton reasoned that the loads imposed on the rear sponsons would be lower than those on the forward step, because any wave encountered at the rear would already have been encountered at the front, thus the forces imposed on the sponsons should be less violent. By contrast, Du Cane judged the loads on the sponsons as the important ones.

Du Cane stated: "I am by no means convinced that this represents a true picture of what might be expected to occur... whilst it is not unreasonable to expect a boat of this nature under adverse conditions to leave the water surface by as much as 6', this can very well happen without involving loads approaching this magnitude."[156]

More tellingly, Du Cane added: "I agree that the acceleration is something like 7,000 ft/sec/sec under these [conditions] but can you honestly believe that this boat would not be able to pass over a 3" wave? We seem to be overdoing the theory a bit. I am not for a moment trying to dispute your mathematics but there is something wrong with a result of 220 g [*author's note: misquote of Railton's 22 g*]!'

Du Cane then went on to explain that he believed 'spill out' on either side of the planing surfaces would "cushion the shock".

In response, Railton emphasised his concerns over the forward step: "I expect to sail on November 30th, and should like to have another confab with you this week, both about the rudder and also the general strength of the structure, after you have had time to check it over. Will phone you tomorrow, Tuesday. *However, it is a fact that, in level running, the unit-loading of the forward step (allowing for the 'shift' of C.G.) is about six times that of each sponson... as for 'spill out', yes, I would agree, if you incorporate some vee into the surfaces, but you seem stuck on wide flat surfaces* [author's emphasis]."[157]

Because of the reversed layout of the hull, altering its direction presented a problem of its own. For stern-mounted steering, two rudders would have been required — one at the rear of each sponson — because a single, centrally mounted rudder would have needed an overly long drop, and therefore its strength would have been reduced. Twin rudders, however, would have caused added drag and would have had to be matched perfectly with a tortuous linkage. The alternative was to fit a single rudder at the aft end of the forward planing surface, which, because it was forward of the centre of resistance, would give 'reverse' operation: *ie*, the rudder moving to starboard would turn the boat to port. This could have been easily solved by gearing the steering box to reverse any inputs, but it would have been difficult to predict just how much control would be available. Too little deflection would make the craft uncontrollable; too much might cause it to capsize.

Du Cane carried out some trials with a forward rudder on the model, but only with small deflections and at high speed.

Railton concurred with the idea, in a note written on 19th November: "I agree with you that, in this special case, a forward rudder is the logical solution. However, I think it is important that the rudder should be as small as possible and consistent with doing the job. I append some notes on how I think this size should be arrived at."[158]

Du Cane responded on 23rd November by saying: "I also have been thinking on these subjects, but am doubtful as to whether the force exerted by the forward rudder can fail to do the necessary. The trouble with the model tests is that we have been playing around with extremely small angles. I can quite easily sort this out on a $1/16$th scale model. I have to come up to London next Thursday (29th) and would like to suggest that you and John have lunch with me. If this is acceptable we can discuss details later."[159]

Railton then wrote a note to Cobb: "Peter has asked if we will be free for a lunch next Thursday the 29th, and if it fits with you I will telephone him as soon as I hear from you. The only straight dope I got from him is on the rudder, not the size, but the position. I have to say it is the easier of the possible solutions but my fear is it is too far forward. Extraordinarily Peter seems to think the faster you go, the bigger the rudder, and as yet I cannot shift him from this. I have recommended that Ken [Taylor, of Thomson & Taylor] carries out the design work for the rudder and fins with Reg [Beauchamp], they'll do a first class job, and it will be right. I find it very queer that he can

SLOW GOING

A Vosper photograph labelled as "model test". The models used for testing on open water were of ⅙ scale and powered by rocket motors. This model appears to have a scale cockpit, something not shown in any of the drawings for model construction. Courtesy of BAE Systems (formerly Vosper Thornycroft)

model test this rudder at this stage but still cannot continue with the vee-form planing surfaces, which in combination with some small fins on the sponsons, would offer superb stability. I fear lunch will see us leaning on the poor chap about this and the forward step!!"[160] Whatever conversation took place over this lunch is not recorded.

Du Cane certainly understood who was designing the rudder and it was not Vosper, as confirmed by his later mention, in a letter dated 15th January 1952, that "Taylor is dealing with rudder and fins."

Meanwhile, Ken Taylor and Reg Beauchamp went to see Railton before his return to the United States and the notes they handed him would have done little to put his mind at ease:

> We have looked into the most obvious sites for the position of the rudder assembly and enclose our drawings accordingly, however, this is the lesser of several evils. Option one would be two rudders mounted on the stern of each sponson. Mechanically (rods and drop arms) vastly complicated unless the use of hydraulics is considered, though difficult to balance? There is a hydraulic pick-up on the Ghost engine but an installation of

this type puts us outside of the weight limits imposed by Vosper.

A single, offset rudder on one sponson reduces the weight to just under the imposed limit, but would require a larger fixed fin on the opposite sponson to counteract it. This would offer a high degree of stability and is an attractive solution, though it may increase the drag numbers considerably and the hydraulics would be by no means easy to accomplish. The drag stress however would create problems unless the mountings were very strong, and it has been suggested the end bulkheads of the sponsons are of wood alone.

The solution being somewhat forced upon us then, is to mount the rudder at the rear of the forward step, and this is what we have drawn up. As it would be mounted forward of the C/G etc, its operation would be reversed but given the construction drawings we have been sent, size, both in depth into the water and the frontal area that it presents to the water, will be fairly critical, hence the rather dagger-like razor blade shown. At speeds of 80 MPH and over, this rudder scheme is adequate for directional adjustment but may create problems in turning the craft at low speeds, though we have suggestions as to how this can be achieved ranging from a drop-in low-speed rudder, to simply turning the craft with a tender between runs.

As the rudder can be fabricated we can have several 'blades' made up for trial if needed and Corlett has specified suitable materials for both the blade, and the blade mount. We are however in two minds about the swivel mounting as we believe, for safety, the rudder should be mounted from both above and in front (we have drawn a single drop pin mounting/hinge), but the lower brace bracket will need to form part of the rear of the forward step as low as possible. As for the sponson fins, these are on the attached sheets. For these we suggest similar flat plates, similar to the rudder blade, this will allow them to be mounted lengthwise on to the inner sides of the sponsons increasing the area they can be affixed without having to rely on the rather flimsy (for this purpose) wooden transoms of each sponson. We are sure <u>you</u> don't need reminding, but the moment around these will increase the loads upon the forward shoe considerably.[161]

Railton arrived back at home in Berkeley, California, in time for Christmas with his wife, Audrey, and children Tim and Sally.

CHAPTER 7
A LINE IN THE SAND

John Cobb's first letter of 1952 to Reid Railton did not carry good news. Dated 6th February, it stated: "I am very grieved to tell you that Vicki [Mrs Cobb] lost her child last Sunday (stillborn), she is very upset, first child. I can't understand it as she was in perfect health all the time and is now for that matter. The doctors cannot offer any reason for it. Have just heard that the king is dead, what a life. Will write about the boat, it's getting on fairly well I think, was down there last week."[162]

While Railton's reply of 9th February to his long-time friend concentrated on offering his sympathies, he made brief reference to their water speed project: "I have heard nothing of any importance from Peter since I left. Not that I expected to, in fact I think it is only fair to leave him strictly alone at this stage. My guess is that the boat won't be in the water until August at earliest, in which case my arrival about July 1st should be in order." His hand-written postscript added: "You are due a visit to Peter I know. Keep an eye that he is testing the vee-ing as agreed. I suggest if Wakefields [Castrol] are putting up some funds that perhaps George [Eyston] should go along too, as he will be project manager once we see the water."[163]

It appears that these were wise words of advice, as part of Cobb's next letter, dated 18 February, indicates:

> Thank you so much for your very nice letter. Vicki is getting along well now and her natural cheerful nature is standing her in good stead, as luck would have it her new Morris Minor arrived last week so she

has something to amuse herself with.

I went down to see Peter [Du Cane] with George [Eyston] and Watson of CCW [C.C. Wakefield, maker of Castrol oil] and had more runs with the model which went quite well with the rudder, with a few degrees of turn on it, of course it crashed in the mud! He is making the sponsons and I fear they will look pretty awful with those flat bottoms. Also the hull is being framed up. Peter talks of completion in June, but I have my doubts, anyhow I will keep you posted.

Regarding the above mention of 'mud', Cobb was referring to the outdoor testing now being done by Vosper with a ⅙-scale model at the nearby Horsea facility on the northern shore of Portsmouth harbour. Horsea was originally two islands that were joined in 1889 to form an elongated lake for torpedo testing but in later years Vosper used this handy stretch of water for evaluating high-speed, large-scale models.

Cobb's letter went on to say that he was about to travel north to investigate where his new jet boat might be run as the original idea of using Coniston Water, as Sir Malcolm Campbell had done, was in some doubt. Cobb first visited Windermere in the Lake District and then went on to Scotland, to Loch Lomond, where Kaye Don had raised the record in *Miss England III* in 1932 and Sir Malcolm had first run *Blue Bird K3* in 1937.

Five days before Cobb's letter, on 13th February, Peter Du Cane wrote to Railton in order to provide the first proper update since Railton's pre-Christmas visit. When matched to Cobb's remarks about the "pretty awful" sponsons with "flat bottoms", it is plain to see that Railton had renewed cause for frustration:

I have just come in from running the 1/6th scale model with the accelerometer on board. Unfortunately the boat ran so well that she ran ashore eventually after about ¾ mile. However, despite the fact that on reaching the beach she went 8ft into the air and pitched into a mud bank, the cantilevers supporting the sponsons were untouched on one side and only slightly distorted in the fore and aft direction on the other.

The accelerometer rather naturally went off the scale which shows that at least 12g was applied in a vertical direction through the C. of G. In reverse this probably has some consolation to us as the structure of the cantilevers in this model is nothing very particular, yet they unquestionably stood up to acceleration in excess of 12g.

The running incidentally was perfection.

I have definitely established on the 1/6th scale model that the rudder functions in the direction that one would anticipate.

Had your friends Watson and George Eyston down with John the other day to lunch and some model running. All appeared to enjoy themselves quite a lot.

Everything seems to be going quite well at the moment. [Ewan] Corlett is helping considerably over material and technical advice on the cantilever structure.[164]

Again, the lack of detailed information, including no reference to the testing of components that Railton felt were important to the project, was cause for another quick-fire, brief, hand-written 'compliments slip' missive to Cobb on 16th February:

John, has Peter not run any vee form tests at all? We had his word. I am also concerned that the hull is being framed up and Peter suggests a June completion. **What has happened to the forward step/rudder support frame** [*author's emphasis*]? Corlett tells me has received no drawings for this, but has for the engine frame/sponson arm ass'y. More anon, when I have sat quietly!![165]

Railton penned a follow-up to Cobb on 24th February, enclosing a copy of a letter written to Du Cane the same day:

I wonder if he [Du Cane] is planning to give Fairey's a finalised model for test. If 'anything happened', it would be rightly regarded as criminal negligence to have omitted such a test. It is no good him saying that the present model doesn't overturn on the water, and is therefore OK because, (1) it HAS overturned, (2) it is not finalised as regards its external contours, and (3) we don't know accurately what speed it is doing. I think you should insist on this being done, and fairly soon at that.

I also want to be absolutely sure that we get the exact information we want, so when the time comes please notify me, so that we can inform [Maurice] Hooper direct of the precise figures which we want to have.[166]

CRUSADER

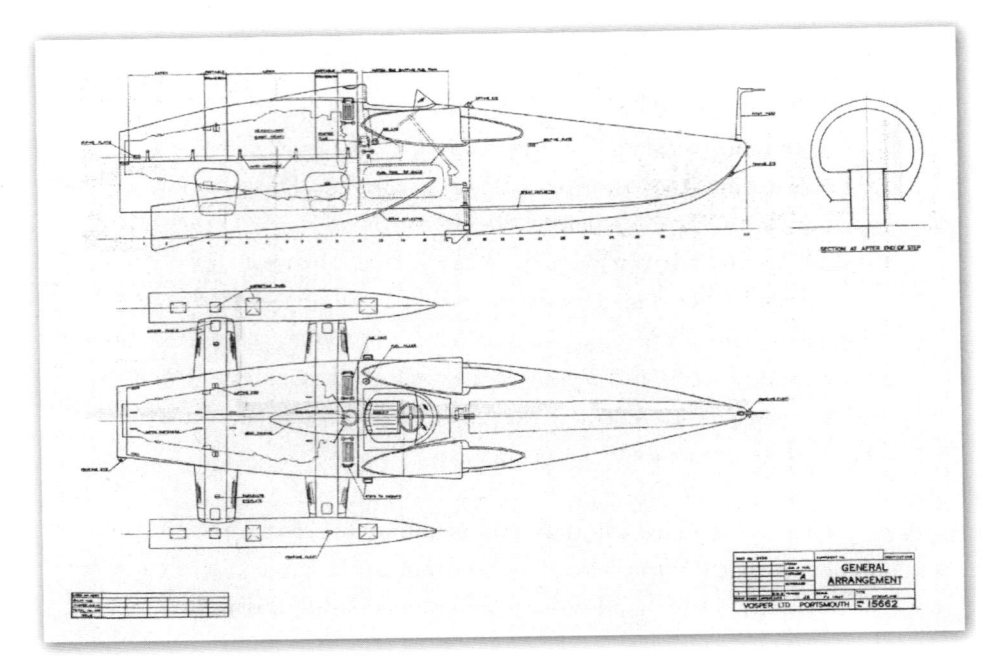

The Vosper 'General Arrangement' drawing. Note should be made of the positions of the lifting eye, cockpit opening, jet engine air intakes, the rudder mounting on the rear of the forward planing wedge, and the position of the wedge. This vertical broken line corresponds to the position of Railton's proposed forward alloy ring bulkhead, which Vosper did not incorporate. Instead there was a plywood bulkhead made from several components as well as a scarf joint in the keel girder. Courtesy of BAE Systems (formerly Vosper Thornycroft)

Hooper, of course, was the engineer in charge of Fairey's wind tunnel in Hayes. It was his task to log and report all findings from every test of *Crusader* models, and it seems that the information was sent only to Vosper.

Railton also went on to remind Cobb of the intention to launch and recover the boat by crane. In his very earliest drawings of the internal alloy frame, he had included two lifting eyes on either side of the centre line of the middle alloy ring frame, and another lifting eye at the centre of the forward alloy ring frame, thereby giving strong triangulated attachment points. As well as Railton's existing concerns about the lack of vee-ing on the planing surfaces and the craft's aerodynamic properties, alarm bells were now also ringing with him about that apparent exclusion of the forward hoop. He expressed his reservations to Du Cane in the letter of 24th February that he had copied to Cobb:

> Thank you for yours of February 13th. I am delighted to hear that you continue to have satisfactory runs with the model: it all seems a very good augury for the future.
>
> There is one thing which I believe was discussed last summer, but which has slipped my memory for some time, and that is the provision of permanent slinging-eyes for picking the boat up with a crane. I wonder if you have remembered about this? I'm afraid I forgot in November.
>
> What I mean is simply two eyes on the fore and one aft of the centre line, one forward of and one aft of the C.G. In the U.S., nobody ever thinks of using a slipway — and we should look such fools if we had a crane available and couldn't easily use it.
>
> At the risk of appearing tiresome, may I urge once again the desirability of providing Fairey's with a 1/6th model with <u>all its external contours true to the final product</u>, for tests on the aerodynamic lift. I won't annoy you by repeating the arguments but you know how I feel.[167]

A brief hand-written note from Railton to Cobb seems to suggest the two had spoken by telephone, and confirms the foregoing concerns:

> I have written to Peter on the three counts. Corlett confirms that it is Peter's plan (already in place now the framing is well advanced) to exclude the forward alloy step/rudder frame, and to launch and retrieve the boat by floating on and off its transport cradle on a slipway. His assumptions being, (1) we would be using Malcolm's slipway at Coniston, (2) the alloy frame was unnecessary as the boat would therefore not be lifted, (3) I had over-calculated the impact shock forces on the forward step. I find this odd as when we discussed this on my last visit he felt the frame was required if we had flat planing surfaces, but that it could be considerably smaller if we had vee'ed surfaces. We now appear to have neither. Corlett could not recall if there was an argument against the mounting of the rudder on the same frame. As for the wind tunnel testing, nothing as yet. I wait to see if Peter addresses these points, or if perhaps I have been too light of foot!![168]

Du Cane's responses, although dated earlier, finally seem to have crossed with Railton's, probably having been sent to the wrong address in America. The first, in reply to Railton's of 24th February, would have done little to please Railton, seemingly confirming what he had heard from Ewan Corlett:

> Your sentiments as expressed in your last paragraph much appreciated.
>
> Reference Fairey's (Hooper), you must not think I am at variance with you over this point. Have already made one visit, on January 18th, at which certain relatively minor modifications were suggested. The matter of the windscreen shape and location were also discussed. The 1/6th scale model in its final state is billed to go up to Hooper for approval, and I feel certain he will manage something for us in the way of actual tunnelling if we leave it to him to organise.
>
> Will go closely into the matter of slinging eyes but must admit up till now have considered only a cradle, which of course would be capable of lifting with a crane given suitable spreaders and slings.[169]

This last paragraph seems to confirm Vosper's unilateral decision to forgo the forward alloy hoop for mounting the forward step and rudder. Part of another letter, written by Du Cane a few days later, indicates that the engine mount/sponson arm frames had not progressed: "Progress reasonable, but still not all material available for cantilever construction, British Aluminium are doing their best, but time is getting along."[170] It must be assumed, therefore, that Vosper's decision to omit Railton's forward alloy hoop was made either because of materials shortage or because it had simply become too late to incorporate it into the engineering build drawings:

Railton then received a reply from Cobb dated 11th March 1952:

> As a matter of fact I have on more than one occasion mentioned to Peter recently the question of returning the model to Fairey's for a final check-up, and I have seen the copy of the letter which he wrote to you on the 3rd March. I am going down there on the 19th instant to see what is going on, and will write again when I return.
>
> I have made no plans yet regarding the lake, but do not quite understand your reference to Loch Lomond, as the North end is quite quiet and free of traffic; I was there many years ago when Kaye Don had *Miss England* up there. However, I do not know what length is available.
>
> P.S. I have no idea why Peter should consider Coniston, Donald has already said it was marginal with the old *Blue Bird*, and that was a

prop! I have never mentioned it to him, so one of his flights of fancy?[171]

For speed, Railton responded with a hand-written note:

> I am beginning to wonder if we are now not stuck in the mire! We have both pushed the issue of getting the model to Fairey's, so to hear there is still no data is perturbing to say the least. Now Corlett informs me he has nothing coming to him either, not even replies to his calling Peter's office. He has informed me there is no problem with the supply of the strong alloy for the frames, so I cannot see why there should be any delay, with all three frames. The ball is in your court again I am afraid, so I see another drive to the south coast. My concern is, that Peter is pushing ahead without taking on board all that we have discussed, and as such I think it prudent to take George [Eyston] with you once again, maybe unannounced to see the state of play?[172]

Cobb wrote to Railton again on 25th March, after his visit to Vosper, and his update suggests that some of Railton's misgivings were probably well-founded:

> I went to Vosper last week and came away slightly encouraged as they seem to be getting on with the job and there is no hold-up for materials. The BA Co [British Aluminium Corporation] are supplying the aluminium F.O.C. Peter states that he is now preparing the model to go to Fairey's so perhaps you will now write to Hooper and state your requirements. The situation there is very delicate according to Peter so I have no doubt you will write in suitable tones.
>
> I am finding Peter difficult regarding the 'sling eyes' as he says he might have to strengthen the hull, and argues that it could be lifted just as easily in the cradle, but I don't see how it could be done if we were on a pier in 6ft of water.
>
> A point I would like your views on is the statement which we shall have to hand to the press before long. Peter's suggestion is to call the boat a 'Vosper-Railton' or 'Railton-Vosper' and state... 'Designed and built by Vosper Ltd under the technical supervision and to meet the requirements of RAR etc etc.
>
> Will you let me have your ideas as I would like to get this matter settled to your satisfaction?[173]

CRUSADER

Yet again, Railton found himself in a position where he felt the need to tread softly, while trying to get things advancing both smoothly and correctly. He wrote the following letter to Cobb on 27th March:

> Re: Faireys. I heard from Peter saying he had found the notes I wrote out on the subject some time ago, and that he was requesting Hooper accordingly, so that's O.K.
>
> Re: the sling eyes. I was afraid he might mumble about 'the strength of the hull', but of course it's too stupid. The main hull members and engine supports are all supposed to withstand 10g, whereas the maximum force that the sling hooks can apply is exactly 1g!!! The trouble is I can hardly write and tell him so without making him look a fool, so I must leave it to you. We certainly ought to have them. His idea is of course that the cradle could be slung submerged off the pier, and the boat floated into position over it, but that is a hopelessly clumsy and outdated method.
>
> I have been thinking quite a lot about this question of a statement to the press etc. If Vosper had been helping you by paying part of the cost, I would have been willing to soft-pedal my part in the deal, as some sort of recognition of their action. As it is, I don't see why we shouldn't tell the exact truth, if only as some sort of recognition to myself for what I have done F.O.C. (I know you will not mind this plain speaking).
>
> So far as the name is concerned, I expect it will be generally known by whatever name you and Wakefield have dreamed up for it. Apart from that, I think it ought to be called a 'Railton-Vosper', <u>not</u> the other way round.
>
> I have roughed out on a separate sheet of paper what might form a basis for the press statement — rather on the lines of what we did for the car years ago. Unless they specifically forbid it, I think we ought to express recognition to Gawn and Co, and the Westcott people [suppliers of the rocket motors for the test models]. The statement has the advantage of being exactly true, and I think contains a fair emphasis on the part Vosper have played, while it certainly doesn't overstate mine.
>
> [*Author's note: Yet again, it is the handwritten postscript that tells the story*] P.S. Regarding the 'sling eyes', the 10g figure is for the hull component only, hence the reason the hull can be lifted by the first and the last of the ***three alloy frames*** [*author's emphasis*].[174]

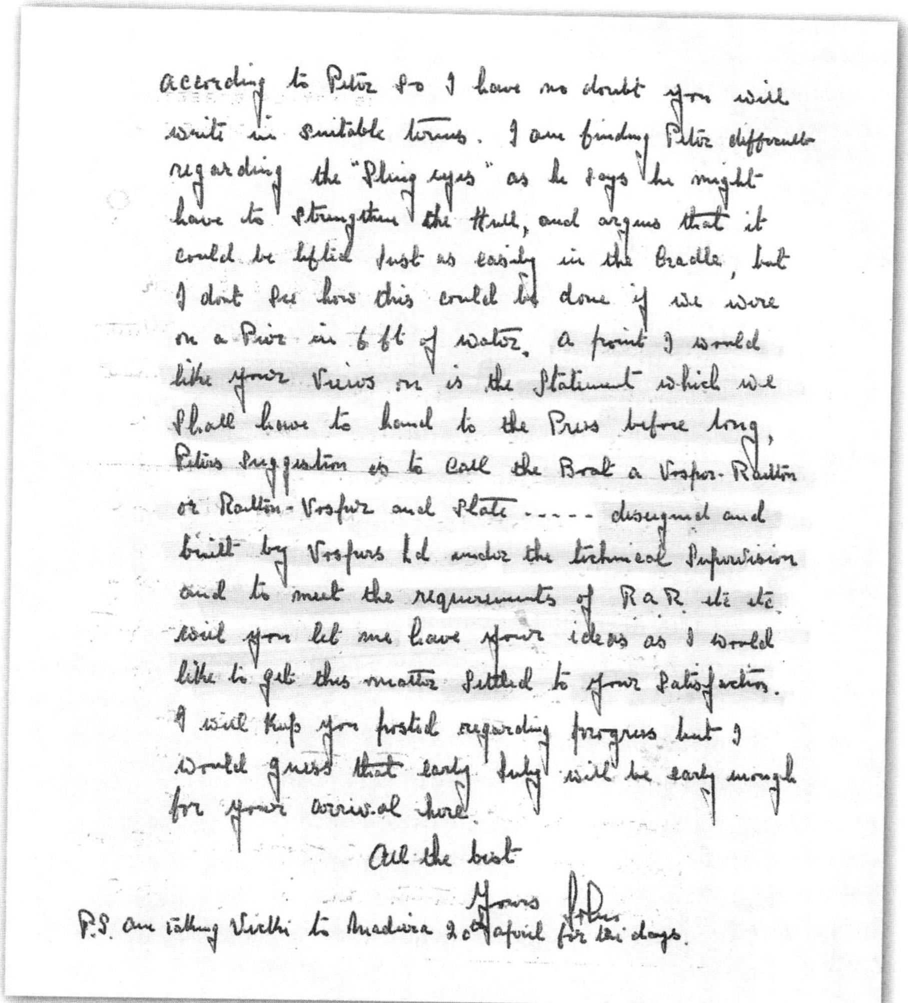

according to Peter so I have no doubt you will write in suitable terms. I am finding Peter difficult regarding the "Sling eyes" as he says he might have to strengthen the Hull, and argues that it could be lifted just as easily in the Cradle, but I dont see how this could be done if we were on a Pier in 6 ft of water. A point I would like your views on is the Statement which we shall have to hand to the Press before long, Peters suggestion is to call the Boat a Vosper-Railton or Railton-Vosper and state ..... designed and built by Vospers Ltd under the technical Supervision and to meet the requirements of RAR etc etc. Will you let me have your ideas as I would like to get this matter settled to your satisfaction. I will keep you posted regarding progress but I would guess that early July will be early enough for your arrival here.

All the best

Yours John

P.S. am taking Vicki to Madeira 20th April for ten days.

The second page of a letter from Cobb to Railton. At this stage framing had begun on the hull, and it is obvious that Vosper had realised that the exclusion of the forward alloy ring bulkhead was going to create a problem with launching *Crusader*, as well as compromising the hull's overall strength. *Railton archive*

This 'PS' clearly shows that Railton had still not been informed categorically that Vosper had decided to omit the forward alloy frame. This is obviously the reason why Du Cane, now clearly faced with a change to how he thought the craft would be launched, informed Cobb that to lift the hull by crane, the forward part of the hull would need strengthening. Whereas Railton, thinking Vosper had followed his design, thought that there was no problem lifting the hull by the alloy internal frame. Of course, this now meant that the stress calculations for that part of the hull were far below what they should have been.

None of this was lost on Lorne Campbell, as he subsequently wrote:

> You have to presume PDC [Peter Du Cane] is referring to strengthening the top of the hull. The 10g is applied to the shoe at the bottom and then, supposedly spread into the hull structure above. This structure could be pretty thin by the time it reaches the top of the deck on the centreline and so — without the ring frame, which I agree should have been incorporated — the top part of the structure would have to be strengthened to feed the hook loads down into the hull. I can't understand why the forward ring frame was left out by PDC. He does seem to have underestimated the loads.[175]

As for the press statement, Cobb gave his brief response to Railton's suggestions in a letter dated 17th April: "I enclose a copy of Peter's ideas and shall be glad if you will let me have it back with any proposed modifications as soon as you can. I notice he has inserted the word 'suggested' instead of 'schemed' [in relation to the design of the boat]."[176]

Railton's return letter of 22nd April was mainly about the press statement. He told Cobb: "I note that Peter has hardly altered the press statement. His substitution of the word 'suggested' is an improvement [*Railton is treading lightly here*], I think, but the bit he has inserted in the last paragraph about 'efficiency of propulsion' is not quite right. I don't suppose it really matters, but if it caught the eye of an expert it would give him a laff. I suggest the alteration shown, and have written to Peter pointing out the reason as tactfully as possible."[177]

As indicated by the following letter to Du Cane about the press statement, dated 21st April, Railton was as tactful as possible whenever he felt the Vosper man had failed to grasp of some of the basic engineering principles involved:

> "John has just sent me your final draft for this press statement thing. In the last paragraph you observe that 'the overall efficiency of propulsion increases as the speed of the boat approaches that of the jet efflux.' While this is of course true on first principles, the maximum efficiency would in practice be reached at a boat speed a good deal less than this. If the speeds were equal, the thrust would be zero, the case being nearly

analogous to that of a propeller and the velocity of its slipstream (see page 202 of your own book).

I would diffidently suggest that you change the sentence to '...while the overall efficiency of propulsion increases rapidly with boat speed', and have marked the copy which John sent me accordingly."[178]

In a hand-written postscript to another letter to Cobb, dated 24th April, Railton was moved to return again to the focus of his concerns.

I fear I may have upset Peter by pointing out he had misquoted himself [in the press statement] from his own book, but I am sure you agree that we have to be right if we aren't to look fools? To be frank, I am tired of treading on eggshells around him, especially on points that are critical to success.

I have to ask you again to glean some more definite answers. As I have been told from outside of Vosper, the sponsons have not been schemed at all as of yet, but the frame, minus the forward ring bulkhead, has, and has been framed up. How is this possible without the final cockpit and jet intake layout having been completed?[179]

It is only with study of period photographs of construction at this time that the last sentence of Railton's postscript reveals its true importance. Looking back to *Blue Bird K3* and *Blue Bird K4*, as well as other contemporary high-speed boats, the internal vertical bulkheads and frames were cut from large single sheets of marine-grade plywood. This ensured that they retained their strength once locked into the hull by horizontal stringers, by remaining in one piece without any joints to weaken the structure at those points. As Vosper had proceeded with construction of the main hull without the forward alloy frame, no areas had been left in the frame for the location of the cockpit and air-intake openings, meaning these would have to be cut into the one-piece frames, creating joints in them. That the frame located at the forward edge of the air intakes lined up perfectly with the rudder mount and the rear of the forward planing step cannot be overlooked, especially as Railton had calculated the stresses for that area thinking there would be an alloy frame. It would be on this wooden frame that Vosper would mount the forward lifting eye.

Photographs of the initial framing of the hull also show that the sponsons were actually being built, even though Du Cane had told Railton that they were yet to be drawn up because model tests still had to be done.

For Cobb's part, he applied himself as normal and had little to report when he wrote on 30th April: "I made a trip to Coniston last Monday, and on arriving there on a beautiful day found the lake like a sheet of glass, and what I was impressed by was the fact that it is nice and straight and just about the right width but of course the facilities are absolutely nil. The boat house Malcolm originally used has been demolished, and even the road approaching it is in such a state of disrepair that the site would be useless. Higher up the lake is the British Railways slipway where *Bluebird* is still resting in a rather pathetic manner."[180]

From Coniston Water, Cobb travelled on to Windermere, where he was offered the use of a slipway, boathouse and docking bay, but he felt the lake was too big, stating, "You could get lost out there."

Less than a week later, on 6th May, Cobb wrote to Railton again: "I visited Vosper last week and found quite some progress has been made, and much to my surprise Peter stated that he had nominated June 15th for completion. I cannot believe this, but at any rate it is something to work on and it looks as though your arrival should be quite well timed, as they will certainly need a lot of guidance on details. I found their idea of a cockpit mock-up somewhat strange and did my best to put them right on various points."[181]

Railton's next letter, dated 11th May, is important for one comment that contradicts many previous accounts of the *Crusader* saga. It concerns the choice of venue for the record attempt: "From what you say, I think Windermere would be the best bet for a real attempt on the record, though I believe both Coniston and Windermere would be uncomfortably short, and that we shall be bound eventually to go elsewhere — certainly to reach anything like 200 mph. Incidentally, from a mere inspection of the map, Loch Ness looks like an obvious place. Have you thought of it?"[182]

Cobb replied on 22nd May. His last sentence simply stated: "I am making enquiries about Loch Ness and will let you know more when you arrive, looking forward to seeing you."[183]

Even though the sponsons were now being constructed, contrary to what Du Cane had told his client, it would still have been perfectly possible for a vee profile to be added to their planing surfaces because they were 'bolt-on' elements and could easily have been remade. Having said that, it was definitely too late to change anything at the craft's bow end, as the frame supporting the very end of the forward planing shoe was now firmly in place — and made of plywood rather than being formed from a ring of alloy construction.

When one closely examines the construction photographs, it is curious that this forward frame ends where it joins the longitudinal bottom runners, and the rear bulkhead of the step is a separate piece, meaning that punch forces — such as impacts with waves at speed — would only be transmitted to the runners via the vertical brackets and a row of screws, reducing the capability of a direct load path into the frame. Of course, Railton had expressed the fact that having a vee section to the planing surfaces would reduce the initial punch force to the shoes, and a one-piece plywood frame would at least transmit much of that force into the hull, although not as effectively as an alloy frame, but Du Cane had seemingly resisted the inclusion of the alloy frame and the vee, to the point where, on 3rd October 1951, Railton had stated his case quite clearly, saying, "I agree that, on the basis of the data which you report, the prudent thing to do is to go ahead with the narrow, flat planing surfaces. It is still inconceivable to me that these flat surfaces will eventually prove to be the best form, in fact it is partly the derision which we shall doubtless encounter when we disclose them which sets me against them. Couldn't you get by with just a little vee at the forward ends?!"[184]

Perhaps one of the things holding back the craft's development had been Peter Du Cane's seeming obsession with eliminating 'porpoising'. This had badly afflicted *Blue Bird K4* after Vosper's jet conversion, and Sir Malcolm Campbell, in his own inimitable fashion, had made Du Cane and his team feel very responsible for the change in the craft's behaviour, regardless of communications from Railton telling Campbell that he was, in effect, wasting his time.

Railton thought he had shown Du Cane a way past this with the veeing of the planing surfaces, although Lorne Campbell disagreed, stating: "I don't think veeing the surfaces did show a way past this — I believe that in Du Cane's mind the models with the vee'd surfaces had all porpoised, but that this was solved when using flat surfaces."[185] It has to be said that if we look onwards to *Bluebird K7*, she also quite successfully used flat planing surfaces.

Yet as we have seen in his August 1952 article in *Motor Boat and Yachting*, Du Cane wrote: "Students of planing forms will perhaps observe the absence of vee or dihedral in the planing surfaces. It would be better in some respects, particularly from the viewpoint of reducing peak loadings, to have had some vee here, but up to the time we had to 'freeze' the design in the interests of production, no really satisfactory behaviour, that is to say freedom from porpoising, had been achieved." Yet in the same article, and previously in

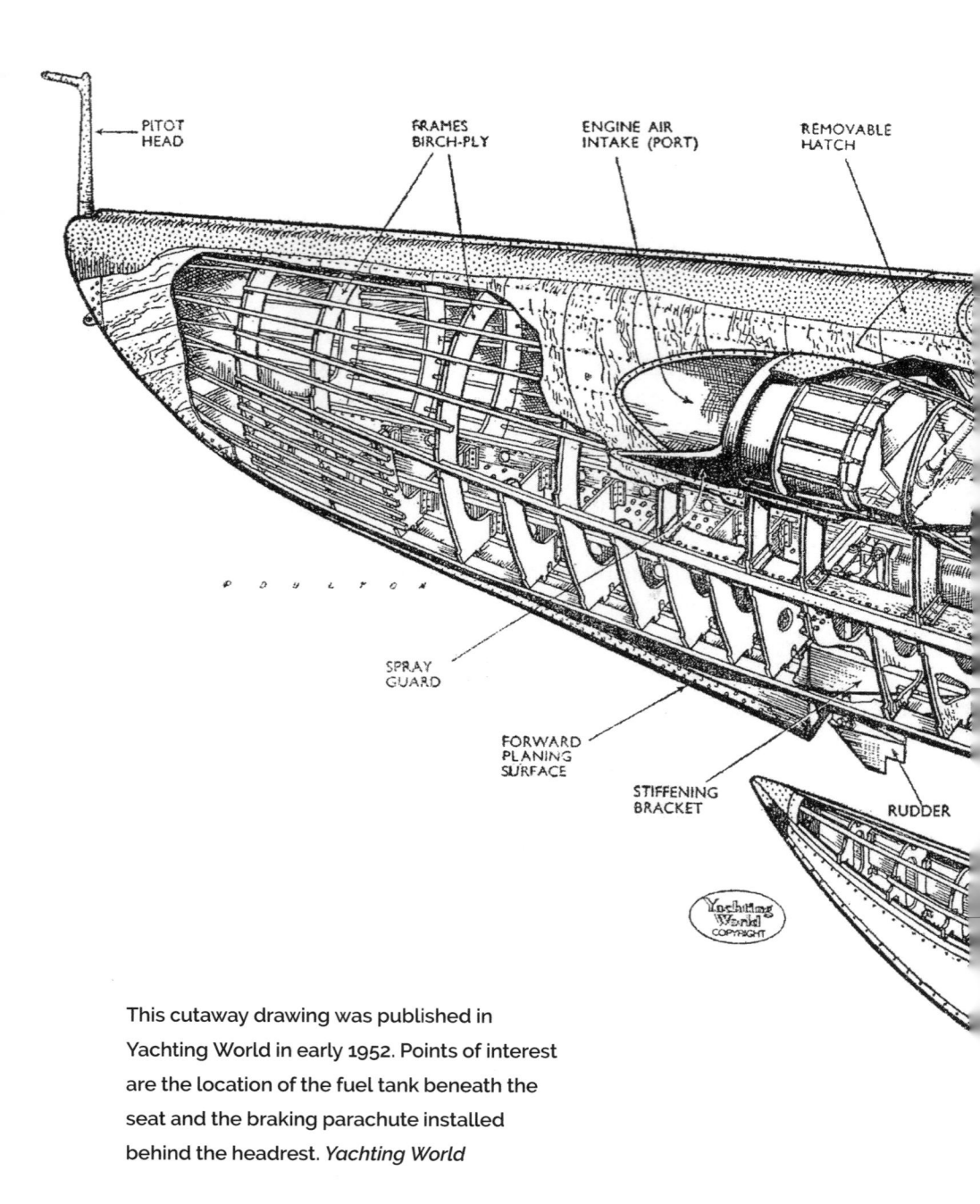

CRUSADER

This cutaway drawing was published in Yachting World in early 1952. Points of interest are the location of the fuel tank beneath the seat and the braking parachute installed behind the headrest. *Yachting World*

A LINE IN THE SAND

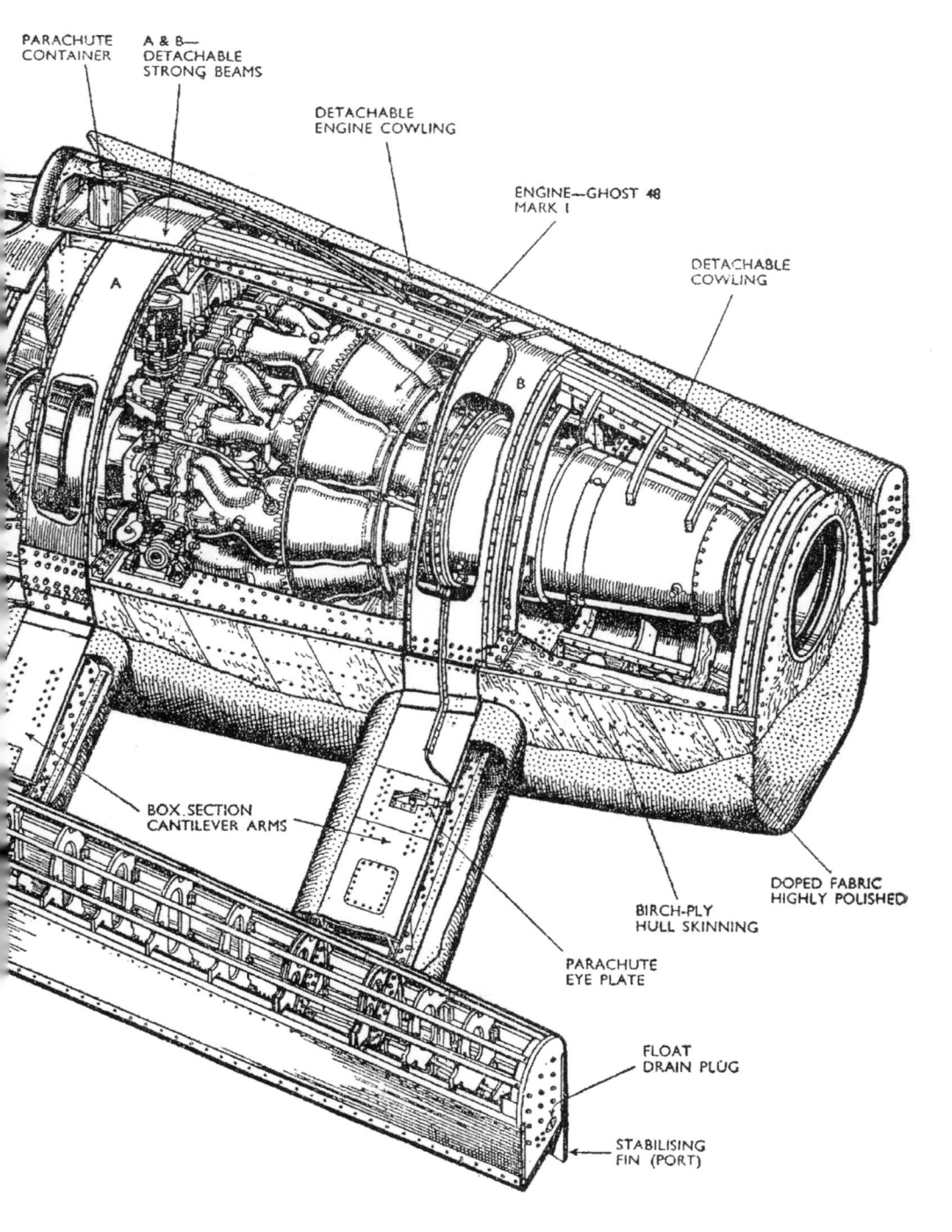

- PARACHUTE CONTAINER
- A & B—DETACHABLE STRONG BEAMS
- DETACHABLE ENGINE COWLING
- ENGINE—GHOST 48 MARK I
- DETACHABLE COWLING
- BOX SECTION CANTILEVER ARMS
- BIRCH-PLY HULL SKINNING
- PARACHUTE EYE PLATE
- DOPED FABRIC HIGHLY POLISHED
- FLOAT DRAIN PLUG
- STABILISING FIN (PORT)

conversation with Railton and Cobb, Du Cane also stated that the "models can only take us some of the way, and we cannot rely on these results in respect of the full-size craft."[186]

Although Du Cane claimed it was too late to include vee'd planing surfaces, clearly it was not. However, as Railton wrote to Cobb: "I fear we have lost the opportunity for the forward step, the crucial one, and there seems little point to include it on the rear surfaces alone, so as Peter asked for my 'sympathy and agreement' [his letter dated 25th September 1951], I have grudgingly had to agree, but it is on paper that I strongly disagree."[187]

Yet in correspondence with some outside the project, Du Cane seemed to be in the same camp as Railton. On 18th August 1952, his friend Arthur Hagg wrote to him after reading the *Motor Boat and Yachting* article: "You have certainly done your best to eliminate the 'unknowns'. I know of course that in operation this 'squib' should go straight as a die but I wondered if you had been able to try in your model the rate of turn at which it becomes unstable laterally. The apparent lack of 'dihedral' on any surface increases my interest."[188]

Du Cane replied to Hagg on 21st August, addressing his friend's concerns about the lack of dihedral or 'vee': "I am quite aware of the limitations imposed by having flat surfaces but frankly could get no satisfactory performance from a model with any dihedral in it **until it was too late** [*author's emphasis*]. I did try the $\frac{1}{16}$th scale model with flat surfaces at 100 m.p.h. with the rudder at 3° and she turned in a controlled arc, admittedly of some pretty large radius. I have briefed Cobb not to try turning too much, but as regards keeping on a straight course I do not think there should be too much difficulty as the fins at the end of the sponsons are enormous comparatively speaking, and in the case of the model kept her straight as a die under all sorts of conditions, and in actual fact, sometimes in surprisingly rough water."[189]

Simply put, there was no reason why the planing surfaces would not have benefited from being 'vee'd'.

Going back a few months, Du Cane made enquiries on 19th March to the Marine Aircraft Experimental Establishment in Felixstowe for an accelerometer, including some technical background and a general arrangement drawing. Superintendent A.G. Smith looked at the matter and also sought the advice of his predecessor and now base commander Wing Commander C.V. Winn DSO, OBE, DFC, who specialised in flying boat and seaplane design and had been part of the High Speed Flight team that had competed in the Schneider Trophy. Between them, Smith and Winn wrote to Vosper querying the flat planing surfaces, stating, "They will induce high-speed pattering and bouncing, **overloading the support structures** [*author's emphasis*]." Their letter also included this decisive instruction from Principal Scientific Officer H.M. Garner: "V them!"[190]

Du Cane responded by stating, 'Although the model performed well on vee'd planing surfaces, it is too late to modify the design as the design was frozen in September 1951."[191] This begs the question of why he was even testing the vee'd surfaces at this stage, although it must have been apparent to him from the Marine Aircraft Experimental Establishment's letter that vee'ing was essential.

It is odd, too, that Du Cane stated that veeing could not be added "as the design was frozen in September 1951", because Railton had persisted in trying to persuade him to incorporate veeing back in May and June 1951, and had been told by Du Cane that "by mid-April some drawings had been done, but they haven't progressed very far". However, the internal agreement to proceed with 'construction drawings' was not signed until 4th July 1951 and none of the necessary drawing work was begun until the beginning of August. It seems inconceivable, therefore, that the adaptation to vee'd planing surfaces could not have been done, especially after master model maker Mr Taylor had reported that the vee'd surfaces had performed exceptionally well.

As Lorne Campbell has observed: "In Du Cane's mind, and probably because of *K4* — the reason he went for flat (zero deadrise) surfaces, was that with some deadrise (vee) he couldn't stop the models from porpoising, at first. What surprises me is that when he says he did manage to get the surfaces with some positive deadrise to work, he didn't modify the full-sized craft — even if the build had started!"[192]

During research for this book, a letter came to light written by Du Cane in 1954 to aerodynamicist Professor P.T. 'Tom' Fink, who was involved with Ken and Lew Norris on the design of Donald Campbell's *Bluebird* and later, in the early 1970s, provided technical advice on wind-tunnel tests and constant design modifications to Ken Warby for his *Spirit of Australia*, the water speed record holder at the time of writing. One paragraph of Du Cane's letter stands out: "One thing which has always puzzled me is that during a very exhaustive series of free model tests with rockets, a number of them in relatively rough water in Portsmouth harbour, we never witnessed any resonant pitching or porpoising, when we had it with vee bottom floats."[193]

And yet Du Cane had still resisted their inclusion — and his continuation of model tests was by now of little use as construction had begun.

CHAPTER 8
CONSTRUCTION BEGINS

Vosper's internal agreement to proceed with 'construction drawings' was signed on 4th July 1951 and chief draughtsman Geoff Brading (with assistance from a Mr Hatch) started work on them at the beginning of August. Vosper's in-house designation for the project was 'Boat 2456'.

As a result of continuing shortages of raw materials in the aftermath of the Second World War, all manufacturing and construction projects in Britain, both military and non-military, required government approval. In response to Vosper's application, Construction Licence SR/452, for the build of a 'High Speed Experimental Craft under 100 gross tons' was issued in late 1951.

Frank Halford of the De Havilland Engine Company had finally managed to come good on his early 'behind-the-scenes' promise of an engine and had persuaded the Ministry of Supply to loan the project a Ghost, which was De Havilland's second turbojet, developed by scaling up its predecessor, the Goblin, and originally known as the Halford H-2. It went on to be used in the Venom and Sea Venom fighter-bombers and in the Comet, the world's first jet airliner. Cobb's actual engine was of the type supplied for use in the Swedish-built SAAB Tunnan, a Ghost 48 Mk I, the unit issued being engine number 2049.

The engine's thrust was 5,000lb and it measured 121in long with a diameter of 53in. To feed it, there was an 85-gallon alloy fuel tank made of AW-3 DTD 213 aluminium alloy. Of *Crusader*'s 6,500lb all-up weight, the engine accounted for 2,218lb.

CONSTRUCTION BEGINS

The hull was a composite construction, of birch plywood and high-tensile aluminium alloy, the need being to keep the weight as low as possible.

Just as construction work started, Peter Crewe, chief hydrodynamicist at Saunders-Roe, wrote to Railton suggesting that Vosper's load figures were "a touch conservative", Crewe having been approached to examine the loads on the planing surfaces based on his work with flying boat hulls. Interestingly, his letter included this comment: "Of course our flying boat hulls present quite a deep V forward of the step, I am surprised you have not, and that the forward step in general appears to widen, not diminish." Extra data on the stresses within the hull was supplied by Vickers Supermarine Ltd and this company seems to have concurred with Railton's figure of 20–22g punch-impact force.

The main hull was built mostly from plywood, supplied by William Mallinson & Sons of Hackney Road, London E2. Two outer skins of thin $\frac{1}{16}$th ply were cross-laid onto longitudinal stringers of spruce, covered with 'doped' fabric and then highly polished. These spruce stringers were in turn glued and 'lugged' (or notched) into vertical birch plywood bulkheads, sawn to shape. The unskinned hull resembled a massive model boat or aircraft.

The 'metalwork' for all highly stressed internal members was of DTD 610B aluminium alloy, as specified by Ewan Corlett. The ring bulkheads were drawn (drawing number 15652) and made more in line with techniques of aircraft construction than of boatbuilding, and were continued out into the mounting arms for the rear sponsons, exactly as Reid Railton had first sketched in mid-1948 with Reg Beauchamp. Where the build differed from the original concept was that there were only two ring bulkheads, not three, and they were not connected horizontally, by beams of similar dimensions. As Corlett had promised "full access to all the materials that you will need", this change may have been due to weight constraints rather than material shortages.

What comes across from surviving documentation is Du Cane's obsession with the two rearward alloy ring bulkheads and sponson mounting arm strength, while seemingly considering the strength of the forward step quite adequate without the alloy ring that Railton had specified. Yet as Railton had stated in November 1951, "in level running, the unit-loading of the forward step (allowing for the 'shift' of C.G.) is about six times that of each sponson..."

The other use of aluminium was for the planing surfaces. From most of Railton's sketches, it appears that he always envisaged these to be aluminium

CRUSADER

Hull framing begins. The vertical plywood bulkheads have been cut from single sheets to retain their strength, as had been specified, but those around the crucial area near the rear of the forward planing shoe would later have to be cut to allow for the cockpit and air intakes, and for a lifting eye to be attached. Although Peter Du Cane had told Railton he would test veeing on the rear sponsons, they were already nearly complete, as can be seen. *Courtesy of BAE Systems (formerly Vosper Thornycroft)*

structures, machined from solid billets with the surfaces incorporated, fixed to the bottom of the forward step and sponsons, as suggested by Corlett during early 'concept' meetings. Peter Du Cane, however, had both step and sponsons drawn up as plywood structures, with the planing surfaces faced with $^5/_{16}$th alloy sheet, a method that presented its own problems.

Du Cane later claimed that he had been persuaded into making this compromise by Corlett, at a meeting on 23rd October 1951, and one has to assume that the change was due to either cost or lack of available materials. However, Corlett had generated pages of data in an attempt to obtain the same strength, panel for panel, with the use of thicker aluminium instead of plywood faced with aluminium, and indeed Du Cane held up construction while testing Corlett's ideas. Eventually, Corlett judged that aluminium sheet alone was unsuitable for the sponson bottoms, unless part of an

CONSTRUCTION BEGINS

Inside the hull, looking forward towards the bow from within the engine bay. Again, no bulkheads have as yet been cut, and the near-complete starboard sponson can been seen to the right. Railton had proposed that the three interconnected alloy ring bulkheads should be made first, and the hull built around them. It is clear here that framing had started even before the alloy ring bulkheads — two of them, not three — had been delivered. *Courtesy of BAE Systems (formerly Vosper Thornycroft)*

all-aluminium structure, machined from a solid billet or — as a very last resort — used to skin a suitably strong flat sheet of plywood, suggesting a minimum ply thickness of 1in. Corlett also stated in writing that the use of aluminium alloy for any reinforcement was *not* to be done, due to the difference in the modulus of elasticity of aluminium and wood, and that this factor should also be considered if Vosper were to face a plywood planing surface with aluminium sheet. He concluded that the 'give' of the materials was so different that any threaded fittings passing through both would be likely to shear.

Whatever the timing during the construction stage, Corlett can hardly have suggested the use of alloy-skinned plywood, as he was very much against the idea and drawings had already been prepared that way. In fact, chief draughtsman Geoff Brading had requested a sample of the planing surface

CRUSADER

In this close-up, the plywood of the bulkheads is clearly visible, as are the plywood block brackets, Vosper at this stage heeding Dr Ewan Corlett's instruction not to mix materials. Courtesy of BAE Systems (formerly Vosper Thornycroft)

aluminium alloy to test and had received it on 9th August, over two and a half months before the meeting between Du Cane and Corlett, and the sample alloy was the same as that specified in the relevant signed-off drawing (number 15661, construction plan sheet 2) dated 28th November.

Boat '2456' shared similarities with both of its predecessors, *Blue Bird K3* (built by Saunders-Roe) and *Blue Bird K4* (built by Vosper). Not only were the new craft's birch ply frames cut from single sheets, but they were attached to what can best be described as a keel girder, made in a similar fashion to that of the *Blue Birds*. The aforementioned drawing, number 15661, shows two longitudinal beams of mixed material (mainly hardwood and birch ply) running the length of the hull, cut to the profile of the hull's bottom, including the angled forward step. The step was attached to the frames with a mixture of notching and lugs. Although the side view on this drawing appears to indicate that the beams were continuous, the plan view shows that there was actually a bolted 'overlap' joint in the area of the cockpit. In between these quite shallow vertical beams, the frames ran through, creating a box effect to the girder, and integrating it into the hull structure, the only exception being on the frame at the rear of the forward shoe. In photographs of construction, as has been mentioned, this appears to have such a drop between the frame height and the step rear bulkhead that it cannot have been a continuous piece (the grain patterns do not match), and that it was joined to the vertical keel members by *aluminium alloy* lugs — which seemed to go against Corlett's advice.

The construction drawings also show that the entire under-surface of the craft was of 1in birch plywood, including the flat planing surfaces, which then should have been also skinned with $3/16$th alloy sheet.

CONSTRUCTION BEGINS

The alloy ring bulkheads: because these were installed after framing had begun, the interconnection between the two rings appears to be by flat sheets at the sides, not the proposed box girders, and there is no interconnection along the floor. Forward, the fuel tank has been fitted, as has the forward engine mount. *Courtesy of BAE Systems (formerly Vosper Thornycroft)*

CRUSADER

Once the frame had been completed, a shell of two layers of cross-laid birch plywood sheet was fixed in place, producing a form of wooden 'monocoque'. The plywood bulkheads within the air intake have yet to be cut through, although in the cockpit area at least three such sections have been removed.
Courtesy of BAE Systems (formerly Vosper Thornycroft)

Many companies supplied equipment and fittings. One was GKN, which provided some 2,000 stainless steel nuts, bolts and washers (2BA and 4BA), although no drawings indicate whether or not these were high-tensile, aircraft-quality fittings. These nuts and bolts were supplemented by 'clench nails' of 12 and 10 gauge, as commonly used in small wooden boats. These look very similar to normal nails but have flat chisel-like tips and once driven through are merely hammered over. Where 'clench nails' were specified in the construction drawings, there is no reference to whether or not they were used with washers. Finally, there were also some areas where 'coach bolts' or 'carriage bolts' were used. These are dome-headed bolts and set screws with a smaller square-section head underneath the dome. As the nut is done up, the square section bites into the wood to prevent it turning as the nut is tightened. The domed head would then be a shallow protrusion above the wood, creating less interference than a normal hexagonal-headed bolt. When Douglas Van Patten was told of their use in the construction, he strongly disapproved.

CONSTRUCTION BEGINS

Engine installation, with the 'test fitting' engine in place. Note that the top half of the forward alloy ring bulkhead unbolts to allow engine installation and removal. It was the sheer diameter of the radial-flow De Havilland Ghost jet that created many of the craft's problems. *Courtesy of BAE Systems (formerly Vosper Thornycroft)*

All in all, the construction of the world's first purpose-built jet hydroplane followed a very established and conservative route. Peter Du Cane rather looked upon it as 'just' another construction project and felt safe in the knowledge that he had Railton advising while keeping with what he knew best. It must be remembered, furthermore, that this was all being done in a post-war Britain where both money and materials were scarce. Indeed, every time Railton visited his home country, John Cobb had quite a search for a car for his use as well as the necessary petrol coupons, and mileage was severely restricted.

An internal memo of 16th July 1952 provides a good indication not just of the complexity of the undertaking but also how at this relatively late stage so much remained to be done, mainly because of delays in the supply of scarce materials. The memo,[194] sent by Geoff Brading to a Mr Rix with a copy to Peter Du Cane, reads as follows:

CRUSADER

Crusader

A list of items to complete the above job is as follows:

Item.	Remarks.
Air ducts to be fitted and bolted in place	Port duct complete from manufacturers. Starboard duct to follow in the course of 3-4 days.
Air duct spray guard	In abeyance.
Cockpit trimming	Seat and trimming being supplied by Messrs. Rumbold.
Dashboard	Complete and being anodised, awaiting erection at boat.
Windscreen	Recommendation regarding manufacture received from I.C.I. representatives. Messrs. Wokington Plastics contacted regarding supply.
Parachute	All necessary material including container in our works.
Head fairing	Templates finished and work in hand.
Heat deflector plate	Information in works and material and fixings in hand.
Fire extinguishing system	All items in store. Ready for piping and mounting in boat as soon as engine hatches are complete.
Forward towing fittings	Items manufactured and ready for fitting.
Pitot head stanchion	Work in hand. Capillary tubes to be run.
Foot throttle control	All material available. Teleflex to cut out controls and install in boat in conjunction with Vosper.
Hand throttle control	Ditto.
Steering column support	Work in hand, to be fitted in conjunction with dashboard.
Accelerometer	Drawing issued, stowage to be provided in float.
Engine cowling top fasteners	Fasteners available. Method of fitting on drawing being issued.
Side frame bracing	Material to hand. Work in progress.

CONSTRUCTION BEGINS

Frame brackets	Work in hand.
Weighing	Arranged for July 21st.
Battery for instruments	Seating to be provided, drawing being issued.
Engine starting battery socket	To be fitted to starboard side of boat, position as arranged.
Cockpit access	Position of footholds to be decided on job. Information issued for manufacturing purposes only.
Pilot's harness	Harness available, fixing to suit seat.
Gripping sockets	Work in hand, to be fitted on boat.
Mooring cleats	Manufactured in alloy, to be replaced by larger plywood cleats.
Float inspection hatches	Drawing in Works. Hatches to be fitted when float is skinned up.
Float drainage	Internal suction hose to be fitted. Pump and hose available.
Boat slinging arrangements	Method of slinging now being designed.
Cockpit hatch	Information in works, reinforcing of cockpit opening to be decided on boat.
Cantilever fairing plates	Work in hand.
Doped fabric	To be completed as soon as air ducts have been finalised.
Fuel piping and vacuum pump pipes	Piping supplied by Palmers. To be installed in boat.
Draining tank and piping	Tank available to be installed and piped up.
Piping and connection for pressurising	Materials available to be installed in boat fuel tanks.
Electrics	Before wiring can be commenced, dash panel to be installed. Fire detector to be wired up. Fuel pressure switch to be fitted to engine. Jet pipe thermo couple leads to be run. Accelerometer lead to be provided and installed. Starter panel for 2nd tender to be manufactured.
Painting	Material available. Position of racing marks to be decided.

CRUSADER

John Cobb's other office: the very basic cockpit and dashboard of *Crusader*. Compared with Cobb's land speed record car, there were relatively few functions to monitor. Railton suggested that alongside the gauges for bearing and jet-pipe temperature, warning lights would be more efficient and easier for the pilot to see.
Courtesy of BAE Systems (formerly Vosper Thornycroft)

On 21st July, *Crusader* was sufficiently complete to be weighed, and Brading communicated the results to Du Cane in an internal memo that listed the weight of 'boat and engine' as 4,608lb and 'items to complete (ex-sling gear)' as 1,633lb, making a total of 6,241lb. This meant that the target maximum running weight of 6,500lb (with pilot, fuel and last-minute additions, including air ducts) was well within reach.

Meanwhile, John Cobb's choice of name had been finally registered. As he had written to his friend Reid Railton, "It was the first name I thought of, and no one has come up with anything I like better, and, ultimately, it's my decision!"[195] In keeping with procedure, the boat was registered with the Marine Motoring Association, which was — by its own definition — 'The National Authority for the Control Of Motor Boat Racing, Cruising & Trials', based at 148 Piccadilly, London W1. The application[196] was made by Peter Du Cane on 25th June 1952 as follows, complete with the required remittance of a postal order for seven shillings and sixpence:

CONSTRUCTION BEGINS

Reid Railton, John Cobb and Peter Du Cane at the London press launch at the Royal Automobile Club, Pall Mall, on 1st July 1952. Interestingly, Cobb is holding one of the earlier rejected model forms with one-piece sponson supports, as the final version was still under test at Fairey's wind tunnel. *Getty Images/ Hulton Archive*

Name of Boat	CRUSADER
Name of Owner	MR. JOHN COBB
Address of Owner	15 ARTHUR STREET, LONDON, E.C.3.
Club	Marine Motoring Association
Type of Boat	Hydroplane
Class	Unlimited
Weight in Lbs	6,300 lbs
Length (L.O.A.)	31'0"
Beam	13'0"
Draught (Aft.)	1'9½"
Engine Type and Number	De Havilland 'Ghost' No. 2049
Year of Manufacture	1951
Number of Cylinders	–
Bore & Stroke	–
C.C. or Cubic Inch	5,000 lbs. Static Thrust

CRUSADER

    Name of Boat Builder   VOSPER LTD.
    Address   PORTSMOUTH

Just two days later, Du Cane received the completed registration[197] for *Crusader*, K6, from the Marine Motoring Association's Honorary Secretary, R.E. Smith:

> I acknowledge receipt of your letter dated 26th June enclosing application form for registration of 'CRUSADER' owned by Mr. John Cobb. I now have pleasure in enclosing Certificate of Registration No. 6.
>
>   In the past all countries Affiliated to the Union of International Motorboating have been allocated a letter denoting nationality, and in the case of Great Britain this has always been 'K'. From the enclosed copy of the U.I.M. Bulletin No. 6, you will see that there has been an alteration and that the letter is now no longer required, it being replaced by the Union Jack in the case of Great Britain.
>
>   'CRUSADER' should therefore have the Union Jack 9" X 6" and underneath the horizontal 8 and No. 6.

In July 1952, the project was announced officially to the public. In all, *Crusader* had cost Cobb just over £15,000 — about £425,000 at today's values.

The 'fitting-up' engine was removed and replaced with engine #2049, and on 8th August *Crusader*'s unpainted hull was towed into daylight for the first time. Vosper, as a company, was on annual leave, but Peter Du Cane and a few others had stayed on. Once anchored to the ground the engine was successfully run up.

That same day, the first powered run was made by Du Cane on the upper reaches of Portsmouth harbour, near the Vosper works at Portchester, to check basic controls and to attempt to establish the speed at which she would start to plane. The conditions were quite rough, even for a normal boat let alone one designed to run in flat calm, so getting onto the plane proved to be impossible. After two short, unsuccessful runs to try to plane, Du Cane halted the exercise.

Reid Railton, who had returned to Britain for the culmination of the project, was present that day and left a message with John Cobb's secretary not long after leaving Vosper, her shorthand translating to, "Engine start was

CONSTRUCTION BEGINS

Preparing for first launch at Vosper's premises at Portchester. The unpainted hull is lifted as Harry Cole supervises from the cockpit. Originally the purpose of this first trial was only to check for leaks, but Peter Du Cane attempted to get *Crusader* to plane and the engine was slightly damaged by the ingress of salt water. *John Bennetts, courtesy of Julie Newton*

Afloat at Portchester as a small crowd looks on. Harry Cole has his head down, looking for water within the still-incomplete hull; portions were still to be painted or covered in aircraft canvas, 'doped' and polished. On the dockside there is a torpedo launching tube. *John Bennetts, courtesy of Julie Newton*

Peter Du Cane returns to the jetty after an unsuccessful attempt to get *Crusader* to plane in unsuitable conditions. Railton had suggested that it was perhaps foolhardy to try the craft in the sea, and so it proved. *John Bennetts, courtesy of Julie Newton*

As *Crusader* passes, the two alloy ring bulkheads are clearly visible on the unpainted stern area. *John Bennetts, courtesy of Julie Newton*

CONSTRUCTION BEGINS

satisfactory, but PDC [Peter Du Cane] insisted on putting her in the water and attempting to plane. Foolish, far too rough."[198] Railton also noted that after the attempted runs the triangular plates that led back from the rear of the forward step to the rudder had slightly buckled. When this was mentioned, he was told it was probably due to water pressure. When *Crusader* finally arrived at Loch Ness, both of these plates had two circular holes cut into them, and later third holes were added.

With these short trials out of the way, the job list became a little shorter. There were very few leaks to deal with, the major one being around the jet pipe. Some strakes were fitted along the forward chine, to try to cut down on spray before the transition onto the plane, and baffles were added to prevent too much water entering the air intakes for the engine. The rest of the hull was covered in aircraft canvas, and then silver-doped, a process by which a plasticised lacquer — nitrocellulose in *Crusader*'s case — tightens and stiffens fabric stretched over airframes, rendering them airtight and weatherproof. The red striping and registration mark were applied by hand.

Crusader was revealed to the world on 22nd August at a press day at the premises of transport contractor Adam Bros in New Malden, Surrey. It was here that the launch dolly and the long-distance trailer needed to take her to Scotland were put together, and she was photographed sitting on the trailer as if she were to leave as soon as the function was over. While she was filmed and photographed, John Cobb gave uncomfortable interviews, with short, sharp answers to the same repeated questions. It was only when asked about safety precautions that he seemed to relax a little, and gave a longer answer: "We won't be using a special life jacket of any sort, reinforced or what have you, just an ordinary 'Mae West' [a Second World War colloquialism for an inflatable life jacket], because if something happens at that speed, I mean you've just had it, there's no possibility of getting out alive. Mr Railton has said to me, that with record-breaking, you go out, you put your foot down and you get away with it, or you go out, you stop halfway and you ask for the pot, and I tend to agree."[199] Both Pathé and Movietone News cut this, his longest answer, from the cinema edit.

As the journalists, cameramen and photographers departed, the sponsons were removed and loaded onto the Austin truck that would tow *Crusader* on her trailer to the Highlands of Scotland, followed by a smaller lorry carrying spares and a Vosper 'jolly boat' (an 18th century name for a ship's boat used mainly to ferry personnel to and from a ship or for other small-scale activities).

CRUSADER

All dressed up with somewhere to go. Now painted and polished, and with the ever-present Harry Cole on board, *Crusader* is craned out to be loaded for her trip to Adams Transport in New Malden, Surrey. *Private collection*

Loaded on her newly painted transport/launch frame, *Crusader* receives some final attention to paintwork at Vosper before embarking on her journey to Scotland. © *Castrol Ltd*

CONSTRUCTION BEGINS

Press day at Adams Transport, New Malden, 22nd August 1952. This was an occasion that Cobb was dreading, privately hoping that he and *Crusader* could go straight to Scotland and "get on with the task in hand". *Getty Images/Hulton Archive*

Peter Du Cane and John Cobb pose for the press. Uncomfortable as he was when thrust into the limelight, Cobb later claimed that he actually quite enjoyed the press day at Adams Transport, although in one of his interviews he was a little too candid about the dangers that lay ahead of him. *Getty Images/Hulton Archive*

CHAPTER 9
ON SCOTTISH WATERS

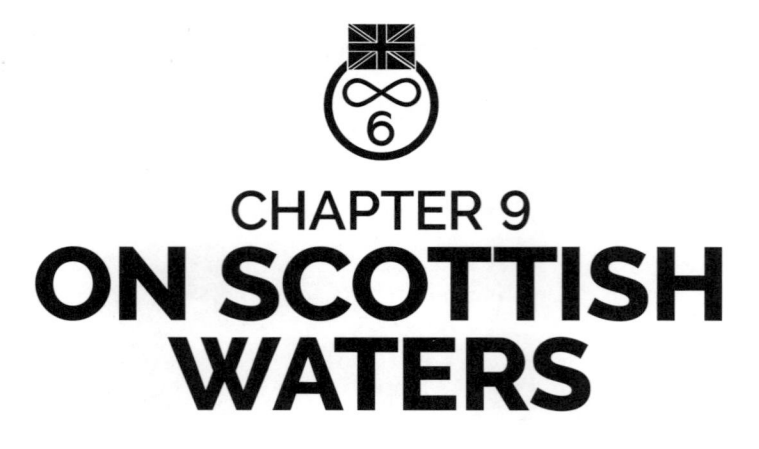

Temple Pier at Drumnadrochit, on the northern shore of Loch Ness, had been chosen as the ideal base for the *Crusader* project, and the pier owner, Alec Menzies, was to become a key figure in proceedings. He recalled:

> C.C. Wakefield & Co. Ltd., of Castrol Oils, as sponsors, approached me with a proposal for the use of Temple Pier as headquarters for the coming event. As this seemed such a wonderful venture to bring to our locality, I was most anxious to bring this, so, as a preliminary step, arrangements were made for Mr Cobb, Mr Railton and Capt. George Eyston to call early in May to inspect the place.
>
> In due course they arrived: Mr John Cobb, tall, of massive build, quiet spoken and unassuming, more like a country doctor than a speed ace that held the World Land Speed Record. In comparison, the designer, Mr Railton was incredibly slight in comparison, and Capt. Eyston was tall and energetic, and a friend of John Cobb, but looked older than his years.[200]

Repairs were made to the old pier. The 'shed' where *Crusader* would be housed was fitted with large doors of highly polished aluminium sheeting on wooden frames, adorned with large Castrol roundels, and a big Castrol banner was hung above the doors. Inside the shed, electricity and a telephone were laid on. A 10-ton capacity Coles crane, weighing some 24 tons, arrived from Sunderland, to be used for engine installation and removal as well as for launching and retrieving *Crusader*. There was a small Ferguson farm tractor for towing her, on her trailer, to and from the pier. A second-hand caravan and a radio van completed the project headquarters.

At the time, an appearance of *Crusader* was described thus: "Such a scene would be marked by the hushed whispers of the expectant watchers on one hand and the roar of the diesel-electric crane on the other. It almost seemed akin to the 'first night' performance of some great masterpiece. The shining doors would open like the parting of a curtain, then slowly but surely *Crusader* would be drawn out from the dark confines into the front of the stage, presenting a magnificent spectacle of silver and crimson, and the scene was set."[201]

Soon after *Crusader*'s arrival at Loch Ness, her sponsons were bolted in place. Vosper employees Hughie Jones and Harry Cole carried out this somewhat awkward task, fitting and tightening the requisite nuts and bolts by means of small access hatches in the sponson fairings, openings that were barely big enough for two hands let alone a pair wielding a spanner. While this was being done, Basil Cronk of Vosper, with John 'Lofty' Bennetts and George Bristow of De Havilland, got on with installing the engine. This had been removed for transportation, for which it had been placed on a special shock-absorbing cradle to protect its bearings.

John Cobb motored up to Scotland with Reid Railton, having left a day or two after *Crusader*'s departure from Portsmouth and arriving on 24th August. Once settled in, Railton and Cobb met with project manager George Eyston. Railton insisted that the team should meet every morning before any runs were made and debrief afterwards, as had been the case at Bonneville during the land-speed attempts.

It was not until Thursday 28th August that Loch Ness and the surrounding hills echoed to the sound of the Ghost engine when it was given its first static test. Already poor weather was playing havoc with the schedule, but Cobb made good use of the time, travelling around and meeting as many local people as he could. Many of them took him to their hearts, appreciative of his promise that no trials would be undertaken on Sundays and of his efforts to give all local children guided tours and the best access for spectating. It was said that Cobb even personally funded the making of small give-away tokens to be handed out to the locals and their children, but, if he did do this, they are so treasured that none has ever been seen.

On Friday 29th August, the tractor, driven by none other than Bert Denly,

CRUSADER

The Crusader's lair: a period postcard of Temple Pier, which was owned and run by Alec Menzies. The site offered the team the chance to work unhindered while allowing the interested and curious to spectate. Author's collection

towed *Crusader* out of the boathouse. Denly was to George Eyston what Leo Villa was to the Campbells, but differed in that he was not only a top-class engineer but also an accomplished racer of motorcycles and cars. Indeed, he had been Eyston's co-driver on many record attempts, and it is said that he once drove Eyston's *Thunderbolt* in a one-way run slightly faster than Eyston himself. When *Crusader* reached the end of the jetty, the Coles crane lowered her into the waters of Loch Ness for the first time. As there were a few 'white horses' out in the loch, the opportunity was used to make sure all the leaks from the Portchester runs had been successfully plugged, which they had. The weather prevented any trials at all during these last days of August, but the time was used usefully to survey the measured mile, Eyston himself rowing out with the bags of cement needed to install the temporary marker posts.

The first date on which the water was suitable for a trial run was Wednesday 3rd September, the very day that Cobb had a pre-arranged commitment in Inverness, 15 miles away. Eyston suggested, therefore, that Peter Du Cane should give *Crusader* her first run on the loch. The modifications to ensure that water would not enter the engine intakes proved to work perfectly, and

ON SCOTTISH WATERS

Nearing the end of her road journey, *Crusader* makes her way through Inverness, with her sponsons carried on the Austin tow truck. The De Havilland Ghost jet engine travelled separately. *Getty Images/Ray Kleboe*

Soon after arrival at Temple Pier, *Crusader* is greeted by John Cobb, who had travelled up to Scotland two days after his craft's departure. This was also a chance for some 'locals' to have a private viewing. *Courtesy of Gordon Menzies*

CRUSADER

Still wearing her transport cover, *Crusader* is pushed into the boathouse/workshop at Temple Pier in preparation for installation of her sponsons. To the left, Railton (black hat) and Cobb (cap) look on. Later, some of the children watching found they could sneak in under the boathouse walls to gaze upon the futuristic craft. *TopFoto*

In the tight confines of the boathouse, Hughie Jones, Basil Cronk, Harry Cole and Bert Denly lift the sponsons into position. *Courtesy of Paul Denly*

While Bill Rees and Ted Cope of radio giant PYE install the 'push to talk' radio system into Crusader, *crowds have already gathered even before the first launch. Near the camera both George Eyston and Peter Du Cane can be seen, and in the background Vicki Cobb can be discerned standing by the 'HQ' caravan door. Ray Kleboe*

it was not long before Du Cane had *Crusader* up on her planing shoes for the first time. Afterwards, he said he was surprised how quickly she accelerated, but his main concern was her tendency to retain speed on a closed throttle. Data during the model tests had shown that the resistance/speed curve became a horizontal line as soon as *Crusader* got up onto the plane, but as some of these tests were done without scale rudders or fins, it was believed that this data would not translate to the full-size hull. After the run, Du Cane is supposed to have told Basil Cronk, "I had to move the rudder from side to side to get the drag up to slow her down, though she didn't seem to move off line, she did slow, but we might need the bigger rudder."[202]

Just as *Crusader* was about to be lifted from the water, Cobb arrived back from Inverness and wasted no time in changing his clothes and climbing into the cockpit for his first run. As he was getting ready, he chatted to Reid Railton, George Eyston, Bert Denly and his friend Denzil Batchelor. Denly remembered being rather surprised by what was said: "George of course was a very good yachtsman, Olympic standard, and was giving John all sorts of advice. John

CRUSADER

In what would become a familiar pose during five weeks at Loch Ness, John Cobb waits for an opportunity to pilot *Crusader*. *Author's collection*

Proudly displaying a Castrol roundel above the wheelhouse, Hugh Patience's *Astrid* prepares to give some of the team a tour of the stretch of Loch Ness where *Crusader* will be run. John Cobb is nearest the camera, while further forward Reid Railton watches the casting off. *Author's collection*

just looked up at him and said, 'But George, your boats have sails, mine has a ruddy great jet!' It was then that someone, I think it was Denzil, asked John what speed he was aiming for on his first run, and if it would be his personal record, that was when George nearly dropped his pipe! 'Oh,' said John, 'I've done about forty-five, I suppose, in an outboard motor.' Fifty-two years old and without any previous experience of really high speeds in a boat, that took some courage I can tell you."[203]

Cobb made five runs at speeds of around 100mph and was hand-timed on the third one at 101.12mph. He made his last return run at taxiing speed and, as he had done at the end of each previous run, skilfully turned *Crusader*, this time in a gentle arc towards the jetty. When the tow craft reached him, he was sitting on the back edge of the cockpit looking quite happy. On being asked how it went, he shouted back, "Wasn't too bad, bit bumpy."[204]

After drying off a little, he went off to the caravan to debrief with Reid Railton, Peter Du Cane, George Eyston, Frank Halford [of De Havilland) and Basil Cronk. As Denly observed, "George wasn't one for talking, he was a 'doer', but on the drive back to our hotel I asked what the state of play was. It was like getting blood from a stone, but he said they had all seemed pretty pleased, the only problems being voiced were from the Vosper chaps, about being light on the throttle and wanting to change the rudder."[205]

Railton had voiced the opinion that Cobb should do some more runs with the craft as she was, to get some miles under his belt, and that it was pointless to change the rudder because Cobb was not having any problems with it. What was apparent was just how much water *Crusader* had taken on when trying to get 'over the hump' and onto her planing shoes. The suction created by the main hull's fairly large, flat under-surface had made the breakaway quite difficult, and the resultant spray had entered the hull through the cockpit and possibly also via the jet exhaust.

By the next day, Thursday 4th September, conditions on the loch had deteriorated, but Cobb still had *Crusader* launched and made one five-mile run, reaching a hand-timed 151.27mph despite the rough water.

Railton, Eyston and Denly stood watching as *Crusader* was craned out, keen to see how much water left the hull after this longer run. Railton, as ever, was noting things down as he saw them, which usually led to a 'to-do list' for someone. On this occasion the lifting was halted when someone noticed a change to the forward step. Denly went back to the tow tractor and switched off the engine before rejoining Railton and Eyston, who had also called over

CRUSADER

Crusader is swung from her cradle, with Hughie Jones on the stern rope and Bill Rees on the bow rope. This photo offers a look at the extra water strakes fitted to improve water flow.
John Bennetts, courtesy of Julie Newton

Harry Cole gives scale to how small *Crusader* actually was. There are three lifting points on the hull: two on the rearward alloy hoop and one forward of the cockpit.
John Bennetts, courtesy of Julie Newton

Finally swung so that she is facing the shore, *Crusader* is ready to be lowered into the water for her first run on Loch Ness. *John Bennetts, courtesy of Julie Newton*

Harry Cole prepares to disconnect *Crusader* from the lifting frame. Already a small crowd, including press photographers, has gathered on and around the pier. *John Bennetts, courtesy of Julie Newton*

CRUSADER

Peter Du Cane prepares to take John 'Lofty' Bennetts for a trip on the loch at sedate speed while Harry Cole appears to be loading one of the marker buoys.
Courtesy of Julie Newton

Du Cane.

Denly recalled: "I lent into the tractor, and flicked the switch, and went back to where they were all standing, bending down looking under the bow. Railton was walking around, like he was comparing the sides, and as he pointed, so the other two would step in and take a closer look. I stood next to George and asked him what they were looking at, and he just pointed. It wasn't much, but you could see the shiny ali' at the back of the step, the upright bits, one side seemed dented, the other like it was bulging out, the reflections in the metal weren't right, like looking at a spoon instead of a mirror. It looked like it had been hit upwards. The shiny flat bottom of the front shoe wasn't flat any more either, it looked wrinkled, and like it had a hole in it. When Cronk appeared, it was too crowded, so I went to sit on the tractor."[206]

At some stage the same day, Vosper decided that the whole hull should be checked, possibly because of the distortion to the front step, or maybe as a matter of course. To assess the hull properly, it was necessary to remove the engine, so any further runs had to be put on hold. Later that evening, Railton, Cobb, Eyston and Denly dined together at their hotel. Du Cane was not with

ON SCOTTISH WATERS

John Cobb prepares for his first run. Remarkably, this was the first time he had actually sat in *Crusader*, having made do with a mock-up at Vosper. He succeeded in surpassing 100mph on his first outing. *John Bennetts, courtesy of Julie Newton*

them as he was staying in Inverness, although he was about to transfer to a much nearer hotel. As Denly recalled, there was only one topic of conversation: "No one seemed too concerned about the damage, and even John said it was a little rough, but Du Cane had told him to go out as it was not as bad as the sea had been when he first tried her, and suggested giving her a 'good hard run', so they were all wondering if it had been such a good idea, as Reid said, 'Designed to run in flat calm John, flat calm!'"[207]

Two days after the last run, Du Cane wrote an interim report for Cobb, as follows:

> As it seems I shall have to go back to Portsmouth for a few days next week I am giving you my interim impressions gained up to date for what they are worth.
>
> As far as we have gone I expect you will agree the performance is generally promising.
>
> The hump can be passed, no porpoising has manifested itself (so far) and the boat would seem to be capable of high speeds.

Because of the limited size of the boathouse, all engine removal and installation had to be conducted outside with the use of the Coles crane, and many hands. *Courtesy of Julie Newton*

Removal of the Ghost was considerably easier than installing it. Here Hughie Jones (in the cockpit) advises fellow Vosper men Harry Cole, Basil Cronk and Geoff Brading as the engine nears mating with its mounts and air intakes, while behind old hand John 'Lofty' Bennetts works alone. *Courtesy of Julie Newton*

Turning is an uncertain manoeuvre and she rides rough in broken water.

Turning may and probably can be improved by fitting a larger rudder, on the other hand if the smaller one appears adequate at the higher speeds and the manoeuvre of turning at the end of the run can be eliminated it is obviously less resistful in its present state and possibly fractionally safer.

On balance and other things being equal I would have said it would be well worth trying the larger rudder at least once.

As regards rough riding I would say the conditions in which you ran on Thursday were for all normal purposes unsuitable for a boat of this class. During my drive along from Glenmoriston I apologised to Victoria and the party with me saying I feared there would be no running.

However as we know I asked you to give her a good 'touseling' which you most gallantly did. The results point to the fact that better conditions should be selected in future and in this I think you are in agreement.

I think the accelerometer trials are important so we can have some idea what the loadings may be in relation to those selected as a design parameter.

I think it very probable that even on Wednesday you did not in fact exceed 4-5 G though on this point it is obviously impossible to be dogmatic.

The long-awaited anti-cyclone is what you require for running but in the meantime when there is any wind worth mentioning the water would appear to be better to the N.E. of Urquhart point in the wider portion.

The question of fuel stowage had been much discussed and I have reasonable confidence that you will not find it necessary to alter the tank, but in any case would sincerely ask that the stowage of more fuel be avoided if possible as it is undesirable from many points of view to increase the mass.

Hoping you will from now forwards become progressively more familiar and confident in the CRUSADER but that you will on no account feel too hurried as a development of this nature needs time and a stage by stage approach. May I conclude by saying how much I admired your handling of a very novel form of vehicle!ature[208]

Du Cane's comments about the rudder are noteworthy and we shall return to these later.

While the engine was being extracted, the parachute braking system was also removed because it was needless weight, having only been installed in the expectation of running on a shorter stretch of water. With the engine out, Cronk clambered around inside the hull checking the numerous fittings and structures. Although all his findings were noted, no record of them can be found in any archive. By that evening, though, Du Cane had informed both Railton and Cobb that he felt the forward step needed a repair, and that it was "in hand".[209]

No time had been wasted in reinstalling the engine, so access to the forward part of the hull was now extremely difficult. The only people able to squeeze into the space below the cockpit and in front of the fuel tank, where the repair was deemed to be necessary, were Vosper's Hughie Jones and two boys, Dennis Cronk and Charles Du Cane, the sons of the senior Vosper men. As Charles Du Cane recalled: "Being a twelve-year-old, I was one of the very few who could slip into the nose section through the cockpit to help Hughie Jones add more ribs strengthening the section that became distorted."[210]

These plywood ribs were attached to the bottom of the frame where it met the longitudinal keel girder, to supplement the alloy lugs, and were knocked into place and fixed through with screws. When Peter Du Cane was asked about the form the repair had taken, Railton is said to have responded with, "I wouldn't go telling John that, is there nothing better that can be done?" To which Du Cane answered, "No, anything else would require proper facilities."[211]

The nature of the damage to *Crusader*'s forward planing shoe has often been referred to, but it seems that Bert Denly was the only one to actually photograph it. There are glimpses on film footage but they are scarce and unclear. Denly observed the triangular plates either side of the rudder had pushed slightly out of line, as if the whole step had moved to one side. Added to this, the aluminium under-surface — the area in contact with the water — was no longer smooth and was buckling into the gaps between the vertical frames in a series of dents running from front to rear. This would suggest that either the thick plywood skinned by the aluminium was breaking up or that the frames holding it were collapsing — or a combination of the two.

Once the engine was fully reconnected and tested, and the remedial work carried out, the weather closed in and *Crusader* did not see the water again

De Havilland's George 'Guy' Bristow watches the first of two static tests of the Ghost engine from the water's edge, so that he could check the jet pipe. One wonders if cupped hands were enough? *Ray Kleboe/Bristow family archive*

until Wednesday 10th September. In the hope that the early-morning water might be calmer, the team assembled before dawn, and *Crusader* was towed out at 5.00am. But a swell quickly developed and conditions did not improve until after lunch. Cobb was able to make two runs, one at over 180mph. That was enough to show the team that they were on the right track — and would have been a record had it been timed according to the regulations and averaged over two runs. From novice to record speed, Cobb had taken a mere 12 high-speed runs. As Denly said, "John was not lacking courage."[212]

The next day, Thursday the 11th, saw the best conditions yet — and the problem of *Crusader* getting onto the plane returned once more. Back in Portsmouth, Vosper had been revisiting the test models in an effort to reduce the quantity of water being shipped during the 'transition' stage. To assist with 'unsticking' the hull, the pragmatic solution at Loch Ness was to use a small speedboat to create a rough patch of water and run *Crusader* through it. On the out-run this tactic worked, allowing Cobb again to reach a speed in excess of 180mph, but after turning for the return run, and without an artificial area of rough water, *Crusader* was reduced to being towed back.

At this time, Peter Du Cane returned to Portsmouth, while in Scotland a press release was issued. Bert Denly: "We could only assume it had been sent from Vosper, I know George knew nothing about it, but the gist was,

CRUSADER

Just before the second static engine test, Peter Du Cane, John Cobb, Reid Railton, Bill Rees (of PYE) and a young Charles Du Cane wait for the engine to spool up. *Getty Images/Ray Kleboe*

Going nowhere fast. There were two static engine tests and this is probably the second one as previously Cobb had watched from the shore and Bristow (standing over the engine) had been on the water's edge. *Getty Images/Ray Kleboe*

there would be no further runs until Du Cane returned to supervise certain modifications. We had no idea what these were, and when John said he might do some slow runs to keep his eye in, Basil Cronk more or less said he wasn't allowed to. It created a bit of an atmosphere."[213]

While the team awaited Du Cane's return, ropes were tied around the hull — at the suggestion of Hughie Jones — to create 'ridges' on its flat underside, in order to help *Crusader* become unstuck more quickly.

Even though the latest run had been in the best conditions to date, it was obvious that the forward step had distorted again and was in need of further bracing, which was undertaken once again with plywood and hardwood ribs. Of course, the more ribs, the more screws — and the less space to put them. Instead, Railton made the suggestion that, instead of mixing the bracing materials — against the advice of British Aluminium's Ewan Corlett — and to save time, perhaps Vosper could make two square frames, or even one cube-like frame, of DTD 610 alloy and use this to reinforce the failing ribs at the rear of the step, utilising the same bolt holes, and thereby pull the area together in a single braced structure, with minimal gain in weight. Basil Cronk agreed but, when Du Cane returned, Hughie Jones was sent into the hull again with Dennis Cronk to add yet more wooden ribbing and screws.

Denly, loitering by the boat and awaiting instructions from Eyston, was present when work commenced inside: "I was standing right next to the *Crusader* as Hughie crawled in, followed by 'Wee Dennis'. It was an awful route in, but at least it was clean. Hughie got comfortable, and reached up for the bits and pieces, including a light bulb in a cage. He was told not to get this too close to anything, as the heat might be enough to set the dope covering alight, something I was told was quite common. So I was talking to the side of the boat, and Hughie was answering, and all of a sudden I heard him say 'Oops' or some such, and he said some of the screws through the extra ribs had broken, sheared in the gap where they went through the ali' into the wood. Railton felt obliged to raise the issue of mixing the materials in the repair and Du Cane said he would speak with Corlett by telephone more or less at once."[214]

Whatever was said on the telephone, Corlett felt the need to put it in writing as well. Writing on 11th September from British Aluminium's headquarters at Norfolk House, St James's Square, London, Corlett addressed his letter to Du Cane at the Station Hotel in Inverness:

I have heard the most interesting things about speeds the *Crusader* has been doing. It certainly looks very encouraging, and whether I can get up to Scotland or not, I look forward to hearing the great news on Tuesday or Wednesday. The purpose of this letter is in connection with the front planing surface.

The modulus of wood is approximately one-tenth that of aluminium and accordingly if an aluminium plate is stiffened by wood members, it is possible for the aluminium plate to be given a permanent deformation before the wood suffers deformation. It does not mean that a wooden plate or sheet in the same position as the aluminium would not suffer damage; in fact, the stiffeners would be severely damaged under the same conditions. What I am getting at is this. That it is not really desirable to stiffen an aluminium sheet with wood, as too much of the load is thrown onto the higher modulus material.

If you have any further trouble with the front planing surface, I think you should fit aluminium alloy stiffening, and I would suggest Z section formed up from DTD 610 sheet on your folding press. I do not anticipate that you will have this trouble, in view of what you said, but if it occurs I am quite certain that this is the way to get over it and not to go on fitting additional wood stiffening.[215]

The same evening as Du Cane telephoned Corlett, Railton contacted Thomson & Taylor's Ken Taylor, who related the conversation the next morning to Reg Beauchamp:

Ken was not prone to getting excited, but he wasn't at his best that morning. Reid had called him from Loch Ness, and had given him a set of measurements, and asked him to draw up and schedule a frame of twelve lengths of Z section DTD 610, to form up into a cube shape. This was to be drilled to match the fixings in the recess of the keel and frames that formed the rear of *Crusader*'s forward step. Ken said that once Reid had told him why, he had suggested some diagonal cross braces. We could have had that made and up at Loch Ness in under two days. What riled Ken however was the rudder. We had drawn up a frame that could take different size 'blades', for easy replacement in the trial stage, and from our figures we knew what size we needed to reduce down to, as the speed increased. Reid had

"Each time *Crusader* returned after a run, people would look at the front planing shoe" — Bert Denly. The original Thomson & Taylor flat rudder is nicely silhouetted in this view. *Getty Images/Ray Kleboe*

CRUSADER

All is not well. As *Crusader* is brought ashore, attention is drawn to the front planing shoe, to see what effect this run has had. *John Bennetts, courtesy of Julie Newton*

As Hughie Jones guides *Crusader*'s hull down onto her cradle, Reid Railton peers intently at the step. *Getty Images/Ray Kleboe*

informed Ken that Vosper had produced a larger cast rudder, that flew in face of the figures we had supplied when Vosper farmed out the job to us.[216]

Whatever, it was too late for this repair and by the time Du Cane had called Corlett, Hughie Jones was out of the hull, and in his hotel.

The speeds reached so far showed that the thinking behind *Crusader* was sound, but it is clear from film footage recorded at the time that she was only really happy in flat-calm conditions. With even the slightest swell, the short 'wheelbase' between the planing points would cause a fore/aft pitch, easily visible by the spurts of white water that could be seen surging forward and then stopping as the forward shoe alternately lifted and was driven in.

The forward step was a matter of concern within the team. Denly recalled that "every time she was lifted up, or moved, or someone simply walked by, they would look down under the side at the step area. The underside was covered in Irish linen, painted silver, to make her smooth, and this was in tatters after each trip out, and needed replacing, but no one looked at that, always the step first."[217]

At the same time as all this was going on, there was a debate about whether the press were correctly reporting how *Crusader* had come about. For his part, Peter Du Cane had written to his friend Max Aitken at the *Daily Express*. Although the letter in the archive is incomplete and apparently not sent from Scotland, it is obvious what its intent was:

> My Dear Max,
>
> In connection with the *Crusader* trials, I hope you will not mind my pointing out that your correspondent up there constantly refers to Railton as the designer of *Crusader*. This despite John Cobb's hand-out dated 1st July…
>
> We are quite content to do our job in reasonable obscurity…
>
> It is however rather a different thing when other individuals are handed the credit for such work…To be fair R.R. [Reid Railton] himself does not claim to be the designer so this is no dig at him.[218]

Du Cane also wrote in similar vein to several other newspapers and publications. Cobb in fact received a letter from Henry Loebl, Chairman of Vosper, asking him to rectify the situation once the record had been taken.

CRUSADER

For Railton it was easy, as he stated himself, "I am not a boat designer, and would have gotten exactly nowhere in these ventures without the skilled cooperation of experts... My only responsibility has been what to build and why, the more difficult problem of how to build it has been left in more capable hands than mine..."[219]

It was understandable that Vosper, as a commercial boatbuilder, wanted to capitalise on its involvement and promote its work, but the word 'design' becomes the subject of semantics.

Even Railton's earliest drawings have a resemblance to what was to become *Crusader*, and throughout the development and test period he sent copies of his drawings and calculations to Vosper at regular intervals. As has been seen from his communications with Cobb, the same cannot be said in return. Perhaps Cobb himself summed it up best in one of his rare interviews:

> Reid is a designer of cars in Detroit, but with a lifetime of experience in out-and-out speed, with Parry Thomas, Malcolm Campbell and many others, both on land and water. It has taken three years to get *Crusader* from Reid's drawing board into the water. He put his finger on one problem some time ago, which confronted him in designing a jet-propelled boat. He told me, "Instead of the thrust coming from beneath the stern — that is, the propeller — it now comes from three or four feet higher and runs in a straight line, horizontally along the boat." And so he set about solving those problems with the assistance of Vosper, and Reid's drawings were transformed into a boat by Commander Peter Du Cane and his team.[220]

It is clear, therefore, that Railton was the originator of the design in concept and general appearance, and that Vosper did the *structural* design. Railton had come up with the principle and had supplied Vosper with drawings and stress calculations done by Thomson & Taylor. Using this guidance, Vosper's drawing office had produced detailed construction/engineering drawings from which the boat could be built, the company itself referring to these drawings as 'Construction Drawings' or 'Structural Design Drawings'.

A more up-to-date parallel would be that of Colin Chapman and his Lotus Formula 1 team. Although an extremely competent designer and engineer, Chapman in his later years would give his design team rough sketches and verbal descriptions of his thoughts, then leave it to them to come up with the

Hughie Jones struggles to check that the sponson mounting bolts are still tight after a run. As he was of quite compact build, he also had the unenviable task of crawling into the hull to effect repairs to the forward step. *Getty Images/Ray Kleboe*

Peter Du Cane turns away after examining the forward step. Part of the damage can be seen in the 'quilting' of the aluminium of the planing surface, where it has been pushed up between the bulkheads, something that would not have been possible if the structure had followed the drawings and had been backed by one-inch plywood. Note the three circular holes in the brace panels just behind Du Cane's shoulders. *Getty Images/Ray Kleboe*

CRUSADER

Peter Du Cane chalks on the shape for Hughie Jones to remove from each of the side brace panels. This enlarged the three original holes into one triangular one, in the hope that water passing through would help Crusader break free of the suction created by the flat hull underside. Getty Images/Ray Kleboe

drawings that would become the Lotus 25, the 49 or the 72. The seed of the design, the concept, was Chapman's, even if the detailed execution was not.

Peter Du Cane wrote an article about *Crusader* in the issue of *Motor Boat and Yachting* dated August 1952, published before any runs took place at Loch Ness. Du Cane quoted the letter of April 1949 in which Railton suggested the reverse three-point arrangement for *Crusader* and at no point in his article did he claim that either he or Vosper had been considering such a layout — and, indeed, why would they have been? Vosper designed and built conventional hulls, made from the same materials that had been used in boat construction for the previous 300 years. Vosper's involvement in the design of *Blue Bird K4* had been to construct a craft to a design based on another company's formula and borrowing heavily on the previous design of *Blue Bird K3* by Fred Cooper and Railton; the installation of a jet engine in *K4* was 'merely' a conversion. The reverse three-point concept presented to Vosper by Railton and his team was like nothing that had been previously attempted, and Du Cane stated himself in the article that Vosper was "endeavouring to translate Railton's idea into a practical job".[221]

There was no mention of this 'debate' when Du Cane returned to Loch

Ness and called a meeting for 12th September. For his part, he reported on the testing that had been done to reduce the amount of water spray being generated and improved methods of getting *Crusader* to go 'over the hump'. In turn, he was brought up to date regarding the forward step. He was told that each high-speed run was compounding the damage, and each repair was taking more material, and more time. It was Du Cane's opinion that the extra ribbing was "robust enough" but that perhaps the method of anchoring it may not be sufficient. Few in the room can have agreed with him.

Another problem was the angle of the rear shoes, which not only provided the lift necessary to plane but also formed part of the 'mechanism' for driving the bow down, to cancel porpoising at the cost of increasing the shock loads on the forward step — loads that, as Railton had told Du Cane on more than one occasion, were over six times those acting on the rear sponsons. Whereas Du Cane thought that simply doing away with the alloy skinning on the rear shoes would solve the problem, Railton did not. It is not difficult to see how Railton could only surmise that the cause of screws breaking inside the forward step and the hull was the relative movement between the alloy skinning and the plywood — the construction technique that he had been told was used —

After runs, debriefs were held in the caravan headquarters at Temple Pier: from left are John Cobb, Reid Railton, Basil Cronk, Peter Du Cane, George Eyston (standing), Frank Halford and George 'Guy' Bristow. This photo would have been staged as no-one outside the core of the project was usually allowed to attend. *Ray Kleboe*

CRUSADER

On the advice of both Alec Menzies and Hugh Patience, early starts were made in the hope of finding calmer water. The newly cut triangular apertures in the side brace panels are clearly visible — and, in a further effort to get Crusader *'unstuck', ropes have been tied under the hull to roughen the water. Private collection*

although in reality it was the quilting of the purely alloy skin pushing up the wooden stiffeners above it that was breaking the screws where the wooden stiffeners could not be adequately fastened to the frames.

The relative positions of the two main players in the design of *Crusader* are hard to reconcile until we see that one of them, Peter Du Cane, knew the design in its entirety whereas the other, Reid Railton, was in the dark about some of the elements. For example, it is not known at what stage Railton learned that the forward alloy ring bulkhead had been dropped from the design, and who decided to exclude it. And as has been seen from the discussions between various parties regarding the forward planing surface, Railton was under the impression, as was Corlett, that this area had been built as per the description given to them, namely from 1 $^{5}/_{16}$-inch sheet plywood, faced with 3/16-inch aluminium alloy.

Railton had been heard to question why this alloy surface should be displaying a series of dents if the plywood underneath was intact. But Vosper drawing 15661 appears to show *only* the $^{3}/_{16}$-inch alloy sheet — there is no mention of any sheet plywood forming the planing surface.

Years later, Hughie Jones recalled how the repairs were made at Loch Ness: "It was tiny in the hull, you had to enter via the cockpit, and over the foot bulkhead, then squeeze past the fuel tank to get to the stern end of the planing area. We had a light in there, and everything had to be passed down a human chain. I had to hammer the ali' back down, to get it as flat as possible, and then hammer down some wooden brace bars, between the side girders. They were a tight, 'interference' fit, horizontal across each section, and then screwed in from outside the girders into the ends. We also replaced an area of the ali', as it had pushed in much further at that point. The second time I suggested we cut flat squares of thick plywood to fit down between the girders and the frames, **a bit like Mr. Brading had drawn, but in sections, not one piece** [*author's emphasis*] and then fit thicker fixing timbers around the top, so it would be more securely fixed, but I was told that there wasn't time..."[222]

This statement seems to confirm that at some stage a decision had been made to follow Ewan Corlett's advice not to mix the materials used in the planing surfaces, but that the thickness of the aluminium alloy sheet had not been increased. This shows not only why the surface was deforming in such a way that the internal structure was visible in the form of ridges between the dents, but why Railton was puzzled by the terms used in Du Cane's letter — he was not aware that the plywood planing surface had not been included.

Regardless, Du Cane offered to take *Crusader* back down to Portsmouth and rectify the problem at Vosper's expense, evidently feeling responsible. It has been said that he more or less begged Cobb to let him take the hull but was overruled.

This is not how Bert Denly remembered it: "Both John [Cobb] and George [Eyston] had been in similar situations before, and weren't prone to taking unnecessary risks. Reid [Railton] was even more pragmatic and wouldn't trust anything he had no human control over, there was no need to risk anything, and Reid had offered an alternative, as had Corlett, and it could have been done in a matter of days, but Du Cane didn't seem to see it, but then John wasn't keen on the four to six weeks Du Cane was asking for either."[223]

CRUSADER

As the sun rises, *Crusader* is towed out by the 'jolly boat' to test the rope theory. The Ghost engine must have been already running as the 'jolly boat' did not have room to carry the starter batteries. *Private collection*

The hull was inspected again and Du Cane drafted a letter to Cobb summarising the situation. Only the draft seems to have survived and the actual letter was probably not sent. In any case, sending it would have been a somewhat pointless exercise when sender and recipient were staying in the same hotel, the Glenmoriston, but perhaps Du Cane wanted Cobb to 'sign off' on its contents. Written on Monday 15th September, the draft reads:

> Following our meeting last Friday at which the position regarding the possibilities of the CRUSADER being in a fit state and able to go for the record in the near future was discussed, I have listened to all the evidence and carried out a survey of the boat and its structure:-
>
> It is clear that provided the boat can be got over the hump in calm conditions without taking excessive water into the turbine intakes or the hull itself, speeds of or above 200 m.p.h. can be achieved.
>
> In my opinion water in the hull is a function of the time taken in the transition stage between displacement and planing. No important hull or other leakages are detectable.

ON SCOTTISH WATERS

Reid Railton and John 'Lofty' Bennetts prepare to go out on the loch in one of the support boats. *Courtesy of Julie Newton*

The structure in its present form is unsatisfactory at the after end of the forward planing surface. The correct course here is in my opinion to replace the existing aluminium sheeting by something stronger as a panel, e.g. thicker and stronger aluminium, steel, or probably best would be plywood faced with aluminium as originally proposed.

There is evidence of fairly considerable movement of the transom relative to the forward 'ring' member. This is probably due to a bending moment set up in the hull structure just abaft the forward ring member. Further longitudinal strength in way of the bottom appears desirable between the two ring members and perhaps in or over the bottom skin.

Other relatively unimportant features are not quite satisfactory such as fairing, fins, etc., but at this stage I do not regard them as vital.

The correct course in order to ensure taking the record this year would be, in my opinion, to return the boat to our shipyard so we can attend properly and quickly to the structural points mentioned.

At the same time, an investigation in conjunction with experts could proceed as to the best remedy for the 'unstick' problem.

This would envisage not more than one month's delay but would in my opinion ensure the boat being returned here in a thoroughly fit state to attack and take the record with confidence this year.

On your earnest representations that this is undesirable from other points of view, I have considered whether it might be worth making a further effort to take the record without returning the boat to Vosper.

I think provided we are allowed to do some interim strengthening and provided it proves possible to get over the hump with the aid of speedboats making use of near perfect conditions there is a reasonable possibility you could capture the record.

~~I must point out, however, that if for any reason you fail at this stage or further damage is caused, I feel sure you will accept the responsibility and all that it involves.~~

I understand you will accept responsibility for this course.[224]

Du Cane was supposed to have struck out the penultimate paragraph, but it is clear that at no time was Cobb asked or instructed to keep his speed down. Indeed, Du Cane stated in the letter, "It is clear that provided the boat can be got over the hump in calm conditions without taking excessive water into the turbine intakes or the hull itself, ***speeds of or above 200 m.p.h. can be achieved*** [author's emphasis]."

In the draft, Du Cane conceded that the structure at the rear of the forward step was "unsatisfactory" and then, rather curiously, went on to state: "The correct course here is in my opinion ***to replace the existing aluminium sheeting by something stronger as a panel*** [author's emphasis], e.g. thicker and stronger aluminium, steel, or probably best would be plywood faced with aluminium ***as originally proposed*** [author's emphasis]." This confirms two things. Firstly, the construction of the planing shoe had not been done according to the original drawing, "as originally proposed". Secondly, the final build had used *only* aluminium sheeting, not "something stronger as a panel" in the form of 1½-inch thick plywood, the lack of which would have accounted for the series of dents that appeared in the aluminium skinning.

It had also become apparent during this latest hull inspection that there were other structural issues, the most significant being that the transom had moved out of alignment with the forward alloy ring bulkhead. Du Cane thought all the problems could be rectified at Vosper's premises and the boat returned to Loch Ness within a month.

John Cobb in full regalia, with self-inflating life jacket and leather helmet with radio headset. Behind his head is a cover that has been taped shut; a braking parachute had been housed here but it was not needed at Loch Ness and had been removed to save weight. *National Motor Museum*

Crusader speeding along in near-perfect conditions. Flat calm was needed to run well but made getting onto the plane more difficult. *Author's collection*

CRUSADER

Homewood bound. A superb panoramic shot of *Crusader* at speed on a return run. *Courtesy of BAE Systems (formerly Vosper Thornycroft)*

Crusader is towed in after a run, watched by local children. In the 'jolly boat' are Harry Cole, George 'Guy' Bristow, John Cobb, Charles Du Cane, Ted Cope and Peter Du Cane. *Getty Images/Ray Kleboe*

Many years later, Frank Lydall, who was an observer/timekeeper at Loch Ness, recalled what he had seen and heard at the time: "At times it was quite fraught, differences of opinion and the like. The work that was carried out on the front planing shoe, including the drilling of holes inside the bulkhead where the planing shoe was... no one gave a reason for it; but it's what they did. I remember seeing a handful of broken screws that had been brought out. They were in a little pile by the cradle, with people poking at them. We heard there had been a letter, that Vosper thought it was OK to run, but Cobb's team wanted this frame made up and put in. There was a definite rift, with John in the middle."[225]

As for why Du Cane's letter to Cobb exists only in draft form, some experts in record breaking have suggested that Du Cane chose not to send it because he did not want to upset his friend, but there is nothing in previous communications to suggest that Cobb regarded Du Cane — and Vosper — as anything other than a partner whom he had paid to turn Railton's concept into reality. Considering the importance a letter of this type might take on in the event of 'it all going wrong', one would have thought that Cobb *and* Railton would have received copies that would still exist in the archives of their correspondence, but no copy has ever been seen.

It has also been claimed that Cobb wrote a reply to Du Cane in which he agreed to keep his speed down to 190mph and absolve Vosper of any blame should there be an accident. This is surely spurious for all the foregoing reasons.

Without Du Cane's draft letter, Cobb could only go on what he had been told by him and by Basil Cronk, namely that *Crusader* was capable of breaking the record — just — in flat-calm conditions, and afterwards Vosper would carry out the suggested modifications prior to a further attempt the following year.

Finally, another question that has to be asked is this: why were more sturdy repairs not carried out at the time of the engine's removal, rather than go through the less satisfactory procedure of having individuals squeeze into the hull to undertake a 'shoring up' exercise *after* the engine had been replaced?

On Monday 15th September, the official timekeepers were summoned and it was announced that from the 18th onwards all runs would be timed.

By Thursday the 18th Arthur Bray and his timing team were all set to go, and on the morning of Friday the 19th large crowds of spectators had amassed around the loch, many brought in on specially laid-on coaches from Inverness. In the meantime, 17 days had passed and no more remedial work had been carried out, not even Railton's and Corlett's strengthening frame, which, once made, would have taken just two days to install.

This first timed sortie was blighted by a blustery cross wind, and although it did not adversely affect the water's surface, it did make *Crusader* difficult to hold on course. Cobb was officially recorded through the outbound mile at an average of 185.576mph, but during the run he had drifted across the loch and, when he turned to make his return run, he lost his full run-up in getting back in line with the course. He recorded a lower speed of 160.714mph, giving a combined average of 173.14mph — tantalisingly close to the record.

It was now that another small piece of the jigsaw sadly fell into place. Arthur Bray remarked that as Cobb had made his runs at some distance from the loch shore, it had been difficult to track *Crusader* between the course markers. He asked if it would be possible to make further runs closer to the shore, to allow for more accurate tracking. Nearer the shore, of course, any wakes could be quickly and strongly reflected back into the course.

The next day, Saturday the 20th, Loch Ness resembled the North Sea, and Cobb, mindful of Railton's observation that from now on all runs should only be considered in flat calm, decided that *Crusader* would stay firmly on land.

Bert Denly recalled: "Even before dawn it looked unlikely there would be any chance of launching the boat. As usual George, Reid, John and I had breakfast. We were chatting about the old Bonneville days, the *Napier-Railton* and *Thunderbolt*, when John said something about how smooth the salt was compared to water. That's when we were told Vosper wanted to try the bigger rudder, which puzzled us as John had had no trouble with the one that was fitted. I know Reid had said the cast one was too big and too heavy, but they wanted to try it. After they left John said, 'Still bloody experimenting, we'd better not lose a crack at the record for this.'"[226]

Even though no record attempt would be made in less-than-perfect conditions, in terms of either the water or the wind above it, Vosper was keen to try its own rudder.

The larger rudder was handed to Alec Menzies, who then had to spend the best part of a night grinding the rough edges and casting marks off in preparation for it to be fitted. It is possible that the rudder had come straight from the manufacturer, so one has to wonder how symmetrically balanced it was.

Crusader had been piloted by only two people: Peter Du Cane and John Cobb. It was Du Cane who wanted to change the rudder for the larger one, based on his experience at the onset of the Loch Ness trials in Cobb's absence,

ON SCOTTISH WATERS

With John Cobb and Arthur Bray looking on at left, Angus Barr, Castrol's press officer, announces from the deck of the *Maureen* the speeds recorded on 19th September. *Getty Images/Ray Kleboe*

when, after running at less than 100mph, he had stated that he had had great trouble altering *Crusader*'s course. As he climbed from the cockpit, he found that Cobb had returned and was eager to have his first run. Du Cane is said to have told him, "I'd wait until we fit the bigger rudder, she's a cow to turn."[227] Yet Cobb went out and did three runs at higher speeds, turning very effectively at the end of each one. Du Cane conceded that maybe Cobb had been right.

There is no written record of what Du Cane thought of the Thomson & Taylor rudder, but his wish to have it replaced by a Vosper-designed one can be surmised from a letter written to him by his chief draughtsman, Geoff Brading, on 26th August. Brading wrote in response to a request made by Du Cane *before Crusader* had even touched the water: "As requested, I am forwarding herewith the Rudder Calculations for the *Crusader*. When calculating the stress on the rudder blade, I have taken the mean thickness of .25' which seems reasonable. Trusting this is all the information you require."[228]

There then followed two pages of calculations for *two* cast rudders, both larger than the Thomson & Taylor version. One was cast in 40-ton nickel-bronze alloy, the other — a longer blade with a larger surface area — in 100-ton. More importantly, both had a far greater frontal area.

CRUSADER

It was also decided that the three holes that had been cut into the two triangular bracing panels at the rear of the forward step should be added to, and so a large triangular hole was cut below the three round holes in each panel. Although not full load-bearing panels, these sections formed part of the mechanism for transmitting punch loads into the frame, but the thinking was that water passing through each of the new triangular holes from the outside, in and under the hull, might have helped with getting the flat underside or the rear of the vertical transom of the forward shoe to unstick. Once on plane, these triangular holes would be above the water unless the front shoe should drop. Eventually the three circular holes and the triangular hole were combined, to increase water flow through the sides, but leaving a very flimsy-looking brace.

Nearly a week passed before conditions were even approaching favourable. On Saturday the 27th, Cobb was asked to check the larger rudder. He was overheard asking Railton if it was a good idea, saying, "It's not flat calm, is it?". The suggestion was made to merely get *Crusader* to plane, and no more. Cobb reverted to running *Crusader* in the centre of the loch and was recorded at 132mph outbound and 90mph on the return. He stated that he did not like the feel of it enough to attempt to go any faster. There is no record of whether or not the forward step was checked again or, if it was, what was found. Thus far, it had required remedial work every time *Crusader* had been on the water.

For the rest of the day, the team was in for more waiting around until a surprise visit from Queen Elizabeth The Queen Mother, returning to Beaufort Castle from the Spean Bridge Commando Memorial ceremony. She had previously been a guest of Cobb's at Brooklands and, in 1949, at Silverstone, where Cobb had demonstrated his land speed car on one engine at the new circuit's International Trophy meeting. From the newsreel footage shot at Loch Ness, Queen Elizabeth seemed to enjoy the visit far more than her Lady in Waiting.

That evening, Frank Lydall took the stopwatch readings to Cobb, whom he found in conversation with Eyston, Railton and Geoff Brading of Vosper. Lydall stood to one side and waited, later recalling: "The forecast for the next day was ideal, but it was Sunday, and Cobb wouldn't be moved, he would not run. Railton had his crossing home booked as a return, and with no alternative sailings was hanging on as long as he could, but he knew Cobb's mind better than any of us. Cobb was telling the Vosper man he didn't like the feel of *Crusader* now with the large rudder, and that the faster he went he said, the more it seemed to rock side to side. The last thing I heard was Railton leaving, he didn't like crowds, so he went off to his room, and Cobb saying he wanted

A happy distraction to weather-watching occurred when Queen Elizabeth the Queen Mother called in to visit. Here, George Eyston offers some explanation to Her Majesty and Vicki Cobb, while John Cobb stands behind. *Courtesy of Gordon Menzies*

Seen in a still from Pathé News footage, John Cobb and George Eyston bid goodbye to Queen Elizabeth as she leaves Temple Pier. *Pathé News*

the rudder swapping back. Then he turned to me and said, "Now then Lydall, what can I do for you?"'[229]

The larger rudder was left in place.

The use of the larger cast rudder is puzzling, and something that concerned Railton, both at the time and for years afterwards. Indeed, even when working with Vosper in 1948 on Sir Malcolm Campbell's *Blue Bird* K4 during its conversion to jet propulsion, Railton had been moved to write: "I also have concerns regarding the surfaces below the waterline. I understand from your comments, that the boat has been difficult to turn, but the size of the rudder that has been installed puzzles me greatly. Surely, as speeds increase, the rudder should become smaller? The drag must be considerable, and the resistance to inputs at speed huge?"[230]

The quiet giants: Reid Railton and John Cobb. *Courtesy of Gordon Menzies*

And, even as late as November 1951, Railton had written to Du Cane about the rudder. While agreeing that the rear of the forward step was really the only logical place for the rudder, he had added: "I agree with you that, in this special case, a forward rudder is the logical solution. However, I think it is important that the rudder should be as small as possible and consistent with doing the job. I append some notes on how I think this size should be arrived at."[231]

These comments were echoed in Railton's communications with Cobb, and even Reg Beauchamp recalls Ken Taylor being told by Railton to reduce the size of the rudder he had drawn up for *Crusader*: "The faster we go, the smaller it can be and the less effort it will take to trim the course. We're going for a straight-line record, not a manoeuvrability award, and it WILL turn however small it is."[232]

The Beauchamp/Taylor rudder was a flat sheet of extremely high-tensile steel, with razor-sharp edges, held in a fabricated frame. It had been tested to ensure it would not deflect, bend or distort as it was dragged through the water, but would snap if it became presented against the direction of travel, thus preventing it acting as a brake. What Vosper had produced was a much larger, traditionally made, cast rudder of considerable mass, greater length and larger frontal aspect. Railton considered it totally unnecessary and far less safe.

Strangely, in private Du Cane appeared to agree, as indicated by this paragraph from the report of 6th September that was reproduced earlier: "Turning may and probably can be improved by fitting a larger rudder, on the other hand if the smaller one appears adequate at the higher speeds and the manoeuvre of turning at the end of the run can be eliminated it is obviously less resistful in its present state and possibly *fractionally safer* [author's emphasis].

R eid Railton had to set off on Sunday 28th September for his pre-booked trans-Atlantic sailing from Southampton to New York two days later on the RMS *Caronia*. Just before his departure he spoke to John Cobb in private for two hours. One can only speculate what was said, but afterwards Bert Denly drove Railton away and recalled learning from him that there would be some timed, controlled runs and an inspection of the hull before any possible attempt on the record, and that even these runs would be well under *Crusader*'s calculated top speed. According to Denly, Railton added that he was not too unhappy about having to depart, knowing that there was a plan, and suggested he would be returning for the record attempt proper.

CHAPTER 10
THE END

Early on the morning of Monday 29th September it seemed as if patience had been rewarded. The conditions looked favourable and the team was on standby at 4.00am. *Crusader* was fuelled with kerosene and prepared. As the sun rose there was a barely perceptible swell, and by 5.30am the support craft — *Astrid*, *Maureen Mhor*, *Karunna* and Vosper's 'jolly boat' — had been sent out onto the loch.

The *Astrid* was owned by third-generation boatman Hugh Patience, who had taken Cobb out in her to survey the waters on his first visit to Loch Ness. The two had spent the best part of a day going from one end of the loch to the other, with Patience pointing out landmarks and explaining how he used these as markers for currents, stream outlets and fishing spots. The two got on well, and Cobb would assent to Patience's sage advice, as he also did with Alec Menzies. Indeed, it was Patience who, when asked by Cobb about the best time to run, had recommended early morning. When Arthur Bray had suggested running *Crusader* nearer the shore, to allow for better timing, Patience had said that Cobb should then make the return run nearer the opposite shore, for "comfort". The configuration of Loch Ness, which averaged a mile wide and was about 23 miles long, meant that even a relatively slow 40ft vessel could create quite substantial wakes, which could be amplified in places where the loch was a little narrower. Indeed, unusual wakes were a common phenomenon, often mistaken for Loch Ness's rumoured monster. They could move from shore to shore, an effect created as much by the shoreline as by the shape of the loch bed, something with which Patience was familiar.

THE END

With *Crusader* tied to her, the *Astrid* took up station early that morning just outside the bay at Urquhart, near the northern shore. The Vosper 'jolly boat' and the *Karunna* went furthest down the loch, to just before the point at which *Crusader* would turn for the return run, and were stationed alongside each other with the *Karunna* nearer the southern shore. At the mid-point of the measured mile was the *Maureen Mhor* (a 70ft motor yacht built in 1925 and more correctly called the *Corrieghoil*), with Arthur Bray, head of the timekeepers, on board, accompanied by as many press reporters as could be accommodated.

Once conditions had been reported as favourable, Cobb gently rowed out to the *Astrid* to board *Crusader*. The timing allowed over an hour for the various washes of the support craft to die down. All support craft were instructed not to return unless specifically told to do so by Cobb himself. By the time Cobb arrived on the *Astrid*, reports were coming back from the *Karunna* that there was "quite a chop" at the south-west end of the course. Hugh Patience, owner of the *Astrid* and a man experienced in the ways of the loch, told Cobb it was likely that these conditions would persist, so Cobb decided that the run would take place after breakfast.

The telecommunications company PYE Electronics had assigned its chief engineer, Ted Cope, to the *Crusader* project, straight from having worked on the first experimental colour television broadcast, the coronation of Queen Elizabeth II to Great Ormond Street Hospital. Together with Bill Rees, Cope had installed all the radio equipment, including that fitted to *Crusader*, and both men had then stayed on to assist the project, becoming the linchpins of all radio communication. Cope recalled: "About 9.30 Mr Cobb came to me and told me to radio out the stand-down order, to say the run would be after breakfast, that the *Maureen* should stay put and make the best of whatever breakfast they had on board, and that a message would go out when he went back out to the *Astrid*. He also said I should tell the smaller boats, not to go 'gunning it', to keep the wash down. I was sitting at the radio in our van, and I always remember he put this big hand on my shoulder and said, 'Thank you so much, then pop off and get a bite of breakfast', but I said I would stay put and monitor the messages about the water, but typically he made sure someone brought me out tea and a bacon sandwich though."[233]

Cobb and his team went back to the Drumnadrochit Hotel and had a leisurely breakfast while reports continued to come in about the water conditions. It was not until just after 11.00am that there was a positive report. Cobb and his wife Vicki went straight to the caravan by Temple Pier while Cope radioed out for everyone to prepare: "I was told, get them all ready and able, 'We won't be faffing about.' I got responses from all but the *Maureen*."[234]

Cobb reappeared from the caravan in his specially made oil skins and life jacket. Looking out over the bay, he was puzzled as to the whereabouts of the *Astrid*, and therefore *Crusader*. Grabbing Alec Menzies as he was walking by, Cobb asked him, "Where have they put the *Crusader*, Mr Menzies?", but Menzies did not know. The two of them walked to the end of the pier and spotted the *Astrid* nearer the shore, having drifted north-east towards Inverness. Satisfied he knew where *Crusader* was, and that he could get to her, Cobb returned to the caravan, where Ted Cope told him: "Everything is ready Mr Cobb, nothing from *Maureen*."[235] Cobb simply replied, "All right, no panic."

Bert Denly was in the caravan with George Eyston, trying to persuade him that he should come out with him in the *Karunna* so that he could get a look at *Crusader* at speed: "I'd just got George to the point of agreeing when John came back in. He ignored us all and went to Vicki as he picked up his radio mask and helmet. He said, 'Well, Vicki, we'd better get going.' She wished him luck and they just said 'bye'. George had gone, and set off in *Karunna*, and the Vosper lads, Hughie [Jones] and Harry [Cole], had already left in the jolly boat, so I jumped into the *Isabelle* to get John out to *Astrid*."[236]

Once he was aboard the *Astrid*, Cobb stood in her wheelhouse alongside Hugh Patience as they towed *Crusader* out of Urquhart Bay to her start position, with Cobb taking a turn at the wheel. Patience quipped, "I don't think this speed is what you're after, sir!" To which Cobb replied, "It's going just lovely, Hughie." Once in position, Cobb handed Patience his spare gloves and goggles, "just in case", and the two shared coffee from Hughie's flask.

As the *Isabelle* reached the *Astrid*, the *Karunna* and the 'jolly boat' set off towards the south-western end of the loch, but as they did so their crews were surprised to see the *Maureen* coming in. As Cobb climbed into the *Astrid*, Patience pointed and said, "Why, that's the *Maureen*." Cobb looked over and, according to Patience, shouted across, "You were told to stay put, you'll have to go back — we're up against it here." Patience recalled that Cobb was, "About as angry as I had ever seen anyone and he was normally a calm, gentle man."[237]

Indeed, Cobb probably could not even be heard by anybody aboard the *Maureen* but he had been irked enough for a very rare public display of emotion. It is also unlikely that the *Maureen* got all the way back to Temple

THE END

Dawn launch, September 29th, and *Crusader* is lowered into the water. Even though the day was not supposed to bring an attempt on the record, there is already a crowd on the jetty, and the roads around Loch Ness were filling up. Caught in the low sunlight is the larger, polished cast rudder. *Author's collection*

Timekeepers set out to their shoreline timing huts — this was the quickest way to reach the huts as they were not readily accessible from the roads. The figure sitting on the stern is *Picture Post* photographer Ray Kleboe, with Philip Mayne alongside. *Author's collection*

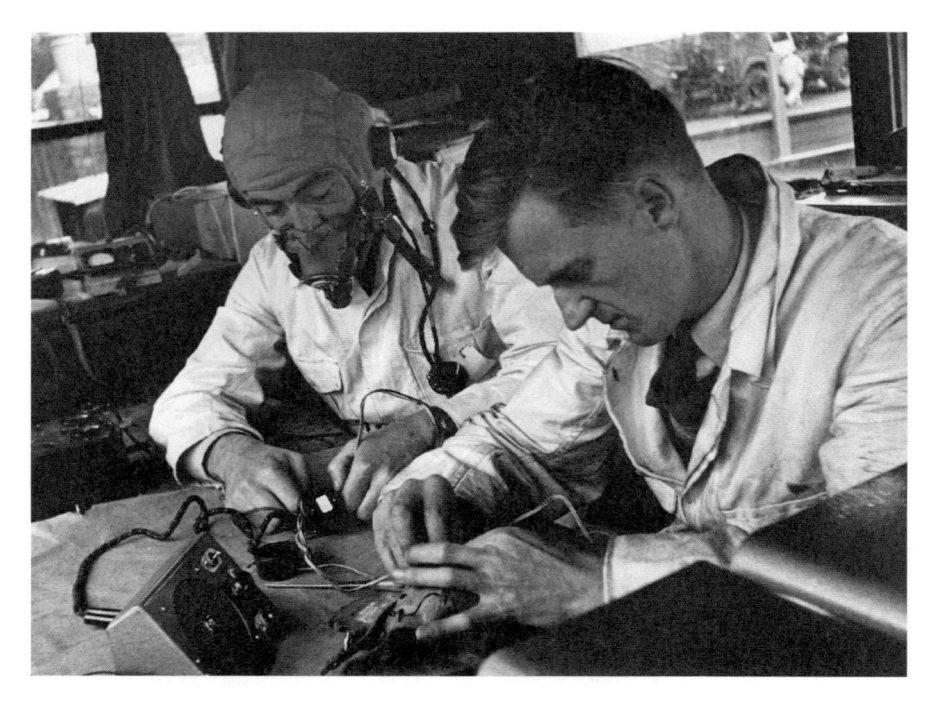

PYE engineers Bill Rees (in headgear) and Ted Cope check part of the project's radio system. *Getty Images/Ray Kleboe*

Pier, and instead it seems that, on seeing *Crusader* being prepared, she was turned around at the Urquhart Castle headland, disturbing the water in a way that might account for *Crusader* carrying instability into the measured mile. Regardless, now instead of the usual hour of waiting for the various washes to settle down, there was a rush to capitalise on the glass-like water. After ten minutes, a green Verey light went up to indicate that the course was ready. Cobb climbed back into the cockpit of *Crusader* as the starter cable was plugged in to connect the batteries aboard the *Astrid*. Cobb edged back into his seat, foregoing yet again the safety belt that had been installed. As he was handed his goggles and gloves, he merely said, "I hope to Christ it's not rough out there." With that, the Ghost jet engine began to spin into life.

Quite why the *Maureen* had returned to Temple Pier is another question that can never be answered. That Ted Cope was unable to get a radio response may be a reason, but Arthur Bray would have known better than to go against the strict instruction to stay at her position. If there were any problems, the drill was to row back to the northern shore from the *Maureen* and go up to the

THE END

As John Cobb answers a question from the press, his wife Vicki looks on. Several of her friends had advised her against being at Loch Ness during the attempt.
Getty Images/Ray Kleboe

road where Peter Du Cane was stationed in the radio car. Once ashore, there would at least be a way of either getting to or communicating with the caravan at Temple Pier.

One possible explanation was that the project was using two separate radio systems. Bill Rees: "Ted [Cope] and I had installed three of our 'DOLPHIN' sets; these went between the timing officials at the marker posts, and one on the *Maureen*. I was the liaison for these, the 'Timing' system, but I was also in contact with Ted who was the kingpin of the main 'Project' system. The 'Project' system went between the timekeepers, the *Maureen*, the *Astrid*, all the support craft, the Vosper radio car and Mr Cobb in *Crusader*. These were kept quite separate."[238]

One has to consider that had someone other than the usual operator on the *Maureen* gained access to the radios and attempted to clear a return on the wrong radio, the lack of response might have led to the conclusion that the radio system had been turned off due to the runs being aborted, and set about returning to base because of it.

As George Eyston assesses water conditions through his binoculars, John Cobb looks thoughtfully across the loch while in debate with Basil Cronk. *Getty Images/ Ray Kleboe*

Regardless of why, return she had, and was now 'rushing' back to the middle of the measured mile. It is also worth noting that after the crash Arthur Bray did indeed radio from the *Maureen*, to call for a doctor and ambulance, so the 'Project' system must have been working.

All this had held up getting under way with the first run, and it was not until 11.49 that *Crusader*'s engine began to spool up and Cobb ran the warm-up procedure. In the radio hut, Ted Cope relayed the last-minute messages: "I had been told that the water on the far [south] side looked better, and relayed this through to '6', Mr Cobb's call sign. All I heard back was 'aye' — it was the last thing I heard from *Crusader*."[239]

THE END

Cobb steps across from *Astrid* to *Crusader* ready to run. The size of the starter battery pack is evident in the foreground. *Getty Images/Ray Kleboe*

CRUSADER

Minutes later *Crusader* appeared from around the headland below Urquhart Castle. Quickly she went up onto her planing shoes. Cobb steered towards the course but, recalling Arthur Bray's comments about being able to observe and time the run, he kept nearer the northern shore, the side where *Maureen* was moored. Having started further back than on previous runs, Cobb worked up to a speed that was very much faster than ever before — in the region of 240mph.

As *Crusader* accelerated, it became obvious all was not well. In a perfect plane a consistent wash was generated from the end of the forward step and the two ends of the sponsons. What could be seen as *Crusader* approached the measured mile was the 'white water' of this wash travelling back and fore along the line of the underside of the step structure, whereas at the very stern the spray pattern remained more constant. This indicated that the bow of the craft was having trouble maintaining its planing attitude.

Then, just as *Crusader* went into the measured mile, three waves could be seen to travel out from the northern shore obliquely at about 45 degrees across her path. The first was struck with what was described as an audible smack and the engine note changed before settling. Then *Crusader* could be seen to rock from sponson to sponson as well as exhibiting progressively worse fore/aft pitch. On hitting the second wave, the bow lifted noticeably before being driven back in by both the thrust line of the jet and the rotation moment about the rear fins. Just as *Crusader* started to settle from this, she encountered the third of the waves, with the bow barely touching water — but once the rear sponsons made firm contact, she appeared to tip in nose first. Somewhere, someone was heard to shout: "He's over." Then silence.

Cobb had been thrown forward out of the cockpit and had bounced across the surface of the water, coming to rest some 150 yards ahead.

The 'jolly boat' started up and rushed to the scene, going first to where the water was covered in myriad small silver shards. As the boat circled, Harry Cole spotted the yellow of Cobb's 'Mae West' lifejacket. Once alongside him, Cole and Hughie Jones pulled Cobb from the water. He had injuries to his chin and right leg, but was inanimate. His lifejacket had burst open and he nearly slipped out of it as he was lifted clear.

Jones: "We struggled to get him out of the water, but we could see it was over. He was just broken, like a bag of broken bones. It was horrible, just horrible."[240]

Cobb's body was laid face down over the engine cover of the 'jolly boat',

THE END

Commander Arthur Bray sets off a Verey light from the deck of the *Maureen*. This flare told those at Temple Pier that conditions were favourable, and scared away errant wildlife. *Getty Images/Ray Kleboe*

which was quite a small craft. Now the *Maureen* had started up and was on her way to the site. Arthur Bray, having been watching through binoculars, radioed base to request a doctor and an ambulance, although those on board the *Maureen* suspected that Cobb was beyond help.

A reporter for *The New York Times* was one of those aboard and his published report of 30th September noted *Crusader*'s rough ride: "The 52-year-old London business man who sought records for the love of speed, had just completed his first run over a measured mile when the boat bounced and then went to pieces in a fraction of a second. Mr. Cobb, who was alone in *Crusader*, was quickly lifted from the water, but his neck was broken. He was dead before he could be brought ashore."[241]

Once alongside the *Maureen*, Bray jumped down into the 'jolly boat' and lifted Cobb's shirt to listen to his back for signs of life. As he stood up, he slowly shook his head. Hughie Jones's face instantly reflected the horror of the moment and he looked on as Bray turned Cobb's head to check his eyes. Clearly his neck had been broken. With that, Harry Cole quickly set off for Temple Pier.

John Cobb is returned ashore after the accident. *Ray Kleboe*

Ted Cope: "I heard from 'AB' [Arthur Bray], on the *Maureen*, and sent someone, anyone, to get a doctor and call an ambulance, so we were thinking Mr Cobb was injured. But there were no telephones free, the press were on them, and it got out before we knew anything. We even heard on site, on the BBC, that he had been injured, someone just cut someone off and snatched the handset and made the call in the end."[242]

The *Isabelle* was dispatched to carry a local doctor, William Macdonald, to meet the 'jolly boat' *en route*. Dr Macdonald confirmed that Cobb was dead.

Bray was first ashore and had to fight his way through a crowd of people who were all now aware there had been an accident and wanted news. As soon as confirmation had come through on the radio from the *Maureen*, Vicki Cobb had been taken to the relative privacy of the caravan. It was here that Bray broke the news of her husband's death.

Out on the loch, a small flotilla recovered fragments of *Crusader*, the biggest only two or three feet square. Frank Lydall described the scene as looking like an autumn fall. Peter Du Cane remembered it as, "A scene of confusion — I saw John Cobb being carried ashore."[243]

THE END

Crusader's funeral pyre. After the accident, the team returned onto the loch and recovered all floating wreckage. At the request of Vicki Cobb, it was burned that evening. *Courtesy of Gordon Menzies*

George Eyston was at the far end of the course and recalled the accident in his book *Safety Last*: "When on station we waited, straining ears and eyes for the evidence which would tell that *Crusader* was on her way towards us. At last, we could just spot a plume of spray in the far distance — the great moment had arrived! The speeding boat was still perhaps six miles away but I could hold her in my glasses. Gradually the plume got more and more pronounced and could be picked up against the shoreline. Then — suddenly — it vanished."[244]

Alec Menzies became a silent witness back at the pier as events unfolded: "Believing *Crusader* had completed her outward run, we waited. Suddenly one of the radio team came running over, shouting, 'There's been an accident, get a doctor, get an ambulance.' Momentarily, I thought he meant a road accident. I knew there was a doctor in attendance on the loch, but it took half an hour to get a clear line to the ambulance service."[245]

At Temple Pier, grown men could be seen crying, while others simply sat in silence. Bert Denly was one of the latter: "I was just stunned. I found something and just sat there without anything I could do or say. Eventually

George [Eyston] found me and sat next to me, and there we sat, in silence. Eventually George said, 'We should get a message to Reid, can you do that? I'll have to tell him what has happened, and this is the one day we didn't have a morning briefing.' He stood, squeezed my shoulder and walked away shaking his head. He was lost."[246]

The absence of the morning briefing that day weighed heavily with Eyston for many years afterwards.

At the end of the jetty the *Maureen* was tied up, her Union Jack at half-mast.

From the British Newsreel sequences taken by Pathé News cameraman Jock Gemmell, and filming done by the BBC, it was obvious that *Crusader* was considerably more unstable than on previous runs, due in part to the increased speed at which she was travelling. Cobb obviously had some indication that something was wrong but had no way of slowing down.

Cobb's body was taken to the Royal Northern Infirmary, Inverness, where a post-mortem examination was carried out. The cause of death was listed as 'Atlanto-occipital dislocation [internal decapitation], subsequent shock from multiple internal injuries'.

On Wednesday 1st October, all the team personnel and close friends of Mr and Mrs Cobb assembled in the tiny chapel of the Infirmary for a service. Outside, crowds stood in silence, many moved to tears. As the cortège departed and passed through the streets of Inverness, the town's flags were at half-mast, while the Provost, magistrates and council members stood bare-headed outside the Town Hall. The cortège briefly halted at Eastgate, where crowds stood in a final farewell, then began its long journey to Surrey. Cobb was laid to rest in the family grave at Christ Church, Esher, on Friday 3rd October. Typically his name is barely visible, engraved on the side of the grave with no mention of his achievements.

The Duke of Richmond, one-time Brooklands racer and creator of the Goodwood motor racing circuit, was moved to write of his friend in the press that evening: "He, as an Englishman, became very great and carrying his fame with an ease quite impossible to anyone with a trace of vanity or conceit. In such a man of action and achievement his intense wisdom could perhaps be expected but it was accompanied by a rare gentleness and sense of humour which made his companionship so much a joy to his friends."[247]

Crusader had actually crossed the timing line for the measured mile, and had averaged 206.894mph, but one run could not count as a record.

CHAPTER 11
WHY?

The definition of an accident is 'an unfortunate occurrence brought about by an unforeseen event, events or circumstances'. Although all accidents start with a single occurrence, a chain of events can sometimes occur. This chain can conclude harmlessly, barely noticed, while other accidents seemingly create a domino effect that ends in calamity.

Investigating an accident requires these chains to be taken apart and examined to see how they interacted, in the hope that the start point becomes clear. To investigate any accident can be a tortuous affair, not always with a clear-cut conclusion. It is sometimes better, therefore, to look at what is known and can be substantiated — and draw the best conclusions we can. If it can be shown to arrive at the same single point, then it can be assumed that it *is* the most likely string of events. *Crusader* was no different.

To that end, we must first look at what we have. There are four films of the accident and some photographs. There are the recollections of those who were there — some involved first-hand, some informed spectators, some merely curious, and some trained observers. Added to this is one other, unique piece of film, the one that came to light just before I completed my manuscript and had not been viewed for 47 years.

To start with, we should examine the observations of those who were there.

Joyce Cronk, the wife of Captain Basil Cronk, Vosper's chief engineer, wrote to Peter Du Cane with her account 20 days after the event, on 19th October.

CRUSADER

> In case it is of any help to you and before my memory dims, I write this account of what I saw of the tragic accident to 'CRUSADER'.
>
> As you know, I was standing by you. I later measured the distance from the first mile post (to ourselves) and found that it was 0.6 of a mile. I did not see the waves beforehand, as, from the moment when I heard my husband over the P.T. ['Public Tannoy'] say, 'he's away', I was watching to see the boat round the point by Urquhart Castle. It was considerably nearer the north shore than on previous occasions and I remember remarking on the beautifully smooth way in which the boat was making the run. It was not until the moment of 'bucketing' that I saw he had hit, what I realised must have been a wave, and then another, the forward part of the boat smacking down on the water with what from sight, must have been considerable impact. As far as I can estimate, the time between the impacts was about half a second and the position of the first bucketing just beyond us about 10 degrees past the direct view across the mile. He seemed to have weathered these and showed no sign of distress at that moment. Immediately afterwards the boat slowed considerably and I next remember the spray rising (about 45 degrees past the direct line) the sound as of an explosion and pieces of debris flying out. From our angle of vision the growing volume of spray hid any further sight of the boat and when it subsided there was nothing to be seen. I am afraid this is a bald account.[248]

Du Cane, who stated that he was standing more or less at the halfway point of the measured mile, released a statement of his own shortly after receiving Joyce Cronk's account.

> On September 6th [*actually it was the 3rd*] I was asked by George Eyston to give the boat a preliminary test to ascertain whether the particular difficulty of getting over the 'hump' (on to her planing shoes) had been overcome. I was gratified to find that after a relatively short period in the transition stage, and provided full thrust was applied, the boat would start to plane relatively easily. Once planing the boat seemed to be very pleasant bearing in mind that I did not at any time pass the 100 mph mark. My report at this stage was to the effect that I did not consider the rudder large enough. However, Mr Cobb then got into the boat and drove her with considerable dash considering it was the first time he

WHY?

Joyce Cronk's handwritten testimony.
Hampshire Archive

From: Mrs. B. Cronk
wife of Captain Basil Cronk

[to retain with other similar letters.]

October 19th
1952

Dear Peter,

In case it is of any help to you, & before my memory dims, I write the account of what I saw of the tragic accident to Crusader.

As you know, I was standing by you. Peter measured the distance from the first mile post, & found that it was .6 of a mile.

I did not see the waves beforehand, as from the moment when I heard my husband over the R.T. say, "He's away" I was watching to see the boat round the point by Urquhart Castle.

It was considerably nearer the north shore than on previous occasions, & I remember remarking on the beautifully smooth way in which the boat was making the run.

It was not until the moment of "bucketing" that I saw he had hit what I realised must have been a wave — & then another, the forward part of the boat smacking down on the water with what — from sight — must have

CRUSADER

Joyce Cronk's handwritten testimony.
Hampshire Archive

been considerable impact.

As far as I can estimate, the time between the impacts was about half a second, & the position of the first bucketing just beyond us — about 10° past the direct view across the mile.

He seemed to have weathered these, & showed no signs of distress at that moment.

Immediately afterwards the boat slowed considerably, & I can't remember the spray seeming (about 45° past the direct line) — the sound as of an explosion — & pieces of debris flying out.

From our angle of view the growing volume of spray hid any further sight of the boat, & when it subsided there was nothing to be seen.

I am afraid this is a bald account.

Yours

Joyce.

WHY?

had been in her and in fact succeeded in turning her after a fashion. A day or two later Mr Cobb took CRUSADER out in rather rough water and gave her a real 'work out' which revealed some minor weakness in the hull. At this stage he did not exceed more than 140-150 mph. A week or ten days later there were very good conditions on the Loch, to the extent that it was almost 'glass calm', which allowed Mr Cobb to run the boat at a much higher speed. His report was that he was easily able to reach 200 mph and in fact a bit more. The acceleration was terrific and full thrust could by no means be applied once this speed was reached. There was however difficulty in glass calm conditions in getting through the transition stage, in fact speedboats had to be used to ruffle the surface of the water at the appropriate place. A thorough investigation of this trouble was proceeding at the time of the accident, and the remedy was in fact discovered, although time had not been available to put this into practice. In the meantime speedboats were used on such occasions to assist in solving this problem.

We had now reached about the middle of September and I asked for an opportunity to go over the hull very carefully and remedy to the best of our ability such defects as had revealed themselves up to date.

It will be appreciated that for a development of this nature, matters had been moving pretty quickly. About a week later I reported to Mr Cobb that I thought there was a reasonable chance of him taking the record this year, provided he was not over ambitious in the matter of speed. From that time forward runs were to be timed officially.

The first occasion on which we timed runs was on Friday, September 19th. On this occasion we were unlucky because although the boat was obviously running very fast, Mr Cobb found himself unable to hold her straight, due, in my opinion, to a 5-6 knot cross wind. He did however even on this occasion accomplish one leg at 185 mph. It was decided to fit a larger rudder, which we had already [had made] against the eventuality that the existing rudder proved too small.

No further opportunity for running occurred until Saturday 27th September, when the weather was far from ideal. However, the CRUSADER was run to prove the efficiency or otherwise of the new rudder. We were gratified to see that the boat was able to turn very convincingly at the end of the run at speeds in the neighbourhood of 100 mph. We now come to Monday 29th September, the day of the final run. Weather in the early morning had proved reasonably promising, but insufficiently good to justify an attempt being made on the record. However, at 11 o'clock conditions improved greatly and all concerned went to their stations again. About 15 minutes were allowed for various washes to die down and the CRUSADER was off. As far as I myself was concerned I was half way along the measured mile on the side of the road about 80/100 feet above the Loch and overlooking it on the North side. This position was of course ideal for observation.

We saw CRUSADER approach, running beautifully at extremely high speed. Unfortunately a system of waves was creeping out into the Loch from the steep shores, and the CRUSADER went into these at extremely high speed with a smack which could be heard from where we were. The bow oscillated violently a few times and the engine was momentarily cut, and then put in again for a short time. From that time forward the boat built up a gradually increasing instability or 'porpoise' culminating in the craft diving in. The exact cause of the trouble I am unable to give, but I am certain that the difficulty occurred after hitting the aforementioned washes as the boat was running perfectly hitherto. From the fact that the average speed of 206.89 mph was recorded, and that for a substantial part of the second half mile, the engine was cut, or at a very much reduced power, I should say that the speed was between 240/250 mph, at the stage when the waves were met. In my opinion more speed than this was available if it could be used.[249]

Curiously, Du Cane stated in this report: "I reported to Mr Cobb that I thought there was a reasonable chance of him taking the record this year, provided he was not over ambitious in the matter of speed" and that the problem with the front step was a "minor weakness". Regarding the comment about speed, Du Cane had certainly stated nothing of the sort in his draft letter of 15th September, and the weakness was anything but minor. Even more curious is the statement regarding the justification of a record attempt, as none was supposed to be taking place when the accident occurred.

Pathé News cameraman Jock Gemmell was using a reasonably powerful telephoto lens. His observations, although seen through a viewfinder (which showed the same magnified image), are useful because he was a trained observer during the Second World War as well as being involved in aerial

WHY?

reconnaissance. He was positioned on the southern shore midway between the *Maureen* and the exit of the measured mile.

> I picked up the boat at quite some distance and it was obvious that *Crusader* was 'pattering' considerably more than on previous occasions, which must have placed a terrific strain on the front skid. She seemed to be bouncing from as soon as I could see her, then suddenly she started to rock sideways and the front bounced a couple times, the last time noticeably higher than before, and when the rear skids hit, she nose dived in, and I followed the pilot as he left the boat, but he was in trouble as soon as he got going.[250]

Opposite Gemmell, on the northern shore, was the camera operated for the BBC. The cameraman, David Low, was a relatively local man from Blairgowrie and had filmed at Loch Ness before, so he knew exactly where to position himself for any attempt on the record. He had been called up into the army in 1941, had served in the famous 'Desert Rats' (Wessex Regiment), had seen action during the D-Day landings, and had been one of the first allied troops to enter Berlin. Once demobbed, he used his wartime observation training by joining Nordisk film studios in Denmark, working as a general assistant in the period 1948–49. On his return to Scotland, he and Robert Riddell-Black set up Templar Films, the backbone of their work being for BBC Scotland Television Newsreel. Low was at Loch Ness in September 1952 at the behest of the BBC. His statement, although recorded many years after the event, is one of great detail, mainly because he was filming from the northern shore, and therefore much closer.

> From watching the previous runs, and being told that Mr Cobb had been asked to run closer to the northern shore, I decided it was worth heaving the equipment down the side to the Loch. I was able to get a good clear shot of the water slightly more than halfway along the course, just beyond the press boat *Maureen*.
>
> I knew any swell would make filming off the boat difficult, so dry land seemed the better bet, even if it meant only getting the latter half of the run, from the *Maureen* on.
>
> We'd been waiting quite some time, when someone shouted down from the road, that they were holding fire until the loch started to behave, so I sat on the grass and opened my flask for a tea.

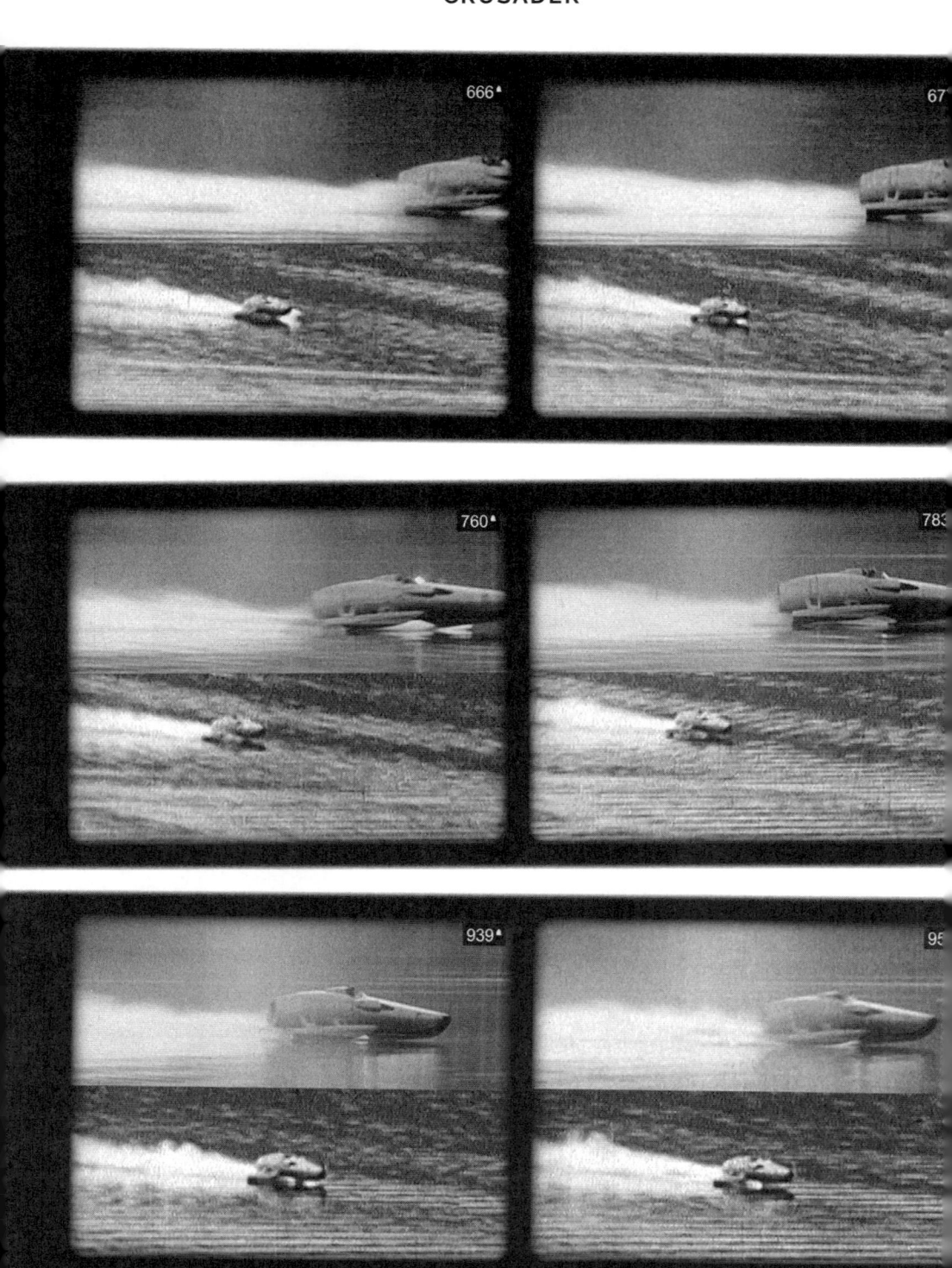

CRUSADER

WHY?

Time-matched stills of the crash from the BBC and Movietone News films. Cobb's head position can be seen to move violently back and fore, and onto the steering wheel. Number 709 shows the rudder clear of the water. Number 939 shows that even though the bow is well down, there is no spray pattern coming from rear of the forward step itself, meaning that the step's height has reduced.
BBC/Movietone News film archives, Leo Villa collection, collated by David De Lara

> After a while I was surprised to see to the *Maureen* starting to turn, and then setting off toward Temple Pier, and looked up to see the radio car had gone, and I thought that maybe they had given it best, and cancelled for the day, but rather than pack up, I opened my sandwiches and decided on breakfast by the loch as the A82 was pretty choked up.
>
> After some time, I was chatting to a couple of young local lads, when one of them said, 'Aye, that's the *Maureen* back', and surely, there she was, closer to the shore than normal, and stopping further to east than normal, and I thought they must have rushed back as there was to be a run after all. So I stood up and got ready as quickly as I could. My viewfinder was a bit out, but I tended to film with one eye to the sight [viewfinder] and the other looking at the target, something we'd been taught in the army.
>
> Very shortly I could hear the whine of the jet engine, and swung the camera around to pick her up as soon as possible, and was quite pleased that *Maureen*'s new position gave me a completely free view.
>
> Quite quickly I saw the boat heading my way, sort of nodding at me. The stern sat really well, but the nose was bobbing considerably. I could see Cobb was having one hell of a ride. One second low in the craft, the next as if he was coming clear out. There was a loud crack, and the boat's whole manner changed, shaking and bouncing, until finally just seconds later she just, well, tipped up at the stern and into spray. Gone, no bang, no explosion, just sudden and awful silence. I am quite certain something broke.[251]

Frank Lydall was one of the timing observers aboard the *Maureen* and as such was also well placed.

> As agreed with Commander Bray, *Crusader* was more on our side of the Loch, and I picked her up through my binoculars as she rounded the head at Urquhart. Though not too well focused, it appeared as if the bow was nodding, though the stern seemed quite firm. As she drew closer, I lowered my binoculars and saw the white water around the bow was changing — it wasn't consistent. Then she shot by and was bouncing terribly. There was a loud crack as she bounced and then disappeared in a cloud of spray. Someone shouted, 'He's in trouble, start her up', and we set off to see what we could do.[252]

WHY?

All Ted Cope could add was this:

> The radio was silent, it was all on 'click to talk' so there wasn't a running commentary from the boat. After she got away, the next thing I heard was, 'He's gone in', I have no idea from who, and then later that 'AB' was asking for a doctor.[253]

Alongside Frank Lydall onboard *Maureen* was 'AB', Lieutenant Commander Arthur Bray, the veteran of many record attempts. His initial statement, made to the police, concurred with Lydall's:

> *Crusader* came towards us travelling faster than I had seen before. It was evident Mr Cobb was having trouble as the bow was clearly bouncing in and out of the water well before the survey point for the mile. As I lowered my binoculars I could see a lot more daylight under the bow, and then a lot of spray. The throttle was kept in, but the rocking continued, and finally she just tipped in nose first.[254]

PYE's Bill Rees was standing next to Bray, fussing over the radios, but had the same clear view:

> I got the double click from Ted [a method of conveying that *Crusader* was 'away' without needless radio chatter] and looked toward the headland that the castle stood on. I could hear the jet, but it took longer than normal to appear and I thought she must have had a longer run up as she was going faster than before. It just wasn't the normal view, usually I could see the wind shield, but the bow was coming up high enough to block it out, then dropping so I could see it. You could hear this whoosh thump, whoosh thump, of the bow hitting the water and rising, which she hadn't done before. Then suddenly, it sort of skipped and thumped, and the engine slowed, like a speedboat jumping a wave, but I couldn't see any waves. Mr Cobb appeared up out of the cockpit, and just as fast, he had disappeared as the thumps started again. I just started to turn to see Bray's reaction, and the boat bounced again and just vanished with a hiss of steam. He was certainly getting a rough ride from the moment I first spotted him.[255]

CRUSADER

There are noticeable differences between the statements made by 'independent' observers and those who were closely connected with the project. In a conversation many years after the event, Joyce Cronk was heard to comment that until she had read Peter Du Cane's first statement, she was not particularly aware of how *Crusader* was running or of any wave formations, but she, like some others who had received a preliminary version of that statement, appears to have used it as an *aide-mémoire*. Du Cane's first statement, published here for the first time, differs from the report that he released later (as quoted earlier in this chapter).

> A series of tests on Loch Ness were carried out by John Cobb and one by myself after arrival up here.
>
> In methodical stages we came to the conclusion the boat was safe for speeds up to about 200 m.p.h. in very good conditions although it was realised that much more speed was available when required or considered safe.
>
> Eventually after minor modifications including a change of rudder we were ready for running for the record by about 25th September 1952.
>
> From that time until Monday 29th, although relatively slow speed trials of 130 m.p.h. were run to test the efficiency of the new rudder, the weather was unsuitable for record breaking.
>
> On Monday 29th September 1952 the organisation stood by for running from 06.30 onwards but was stood down by Mr. Cobb for breakfast at about 09.30 as the weather was unsuitable.
>
> About 11.15 weather had much improved and again the organisation was hastily summoned and the CRUSADER prepared again for running.
>
> This involved a good deal of running of launches from one end of the 5 miles course to the other and by the time they were all in position the CRUSADER was ready to start.
>
> In conjunction with Commander Bray who represented the international body we agreed conditions were suitable but we gave about ten minutes for the water to settle down.
>
> As far as could be seen the Loch was perfect but I did pass a message which have subsequently heard was definitely delivered to John Cobb to the effect that conditions to the South were better than to the North.
>
> After a few minutes I heard on the radio that the CRUSADER was off and shortly thereafter she appeared past Urquhart Castle.

WHY?

Statement of Commander Peter Du Cane,
representing designers and builders of CRUSADER.

1952

A series of tests on Loch Ness were carried out by John Cobb and one by myself after arrival up here.

In methodical stages we came to the conclusion the boat was safe for speeds up to about 200 m.p.h. in very good conditions although it was realised that much more speed was available when required or considered safe.

Eventually after minor modifications including a change of rudder we were ready for running for the record by about 25th September 1952.

From that time until Monday 29th, although relatively slow speed trials of 130 m.p.h. were run to test the efficiency of the new rudder, the weather was unsuitable for record breaking.

On Monday 29th September 1952 the organisation stood by for running from 06.30 onwards but was stood down by Mr. Cobb for breakfast at about 09.30 as the weather was unsuitable.

About 11.15 weather had much improved and the organisation was hastily summoned and the 'CRUSADER' prepared again for running.

This involved a good deal of running of launches from one end of the 5 miles course to the other and by the time they were all in position the CRUSADER was ready to start.

In conjunction with Commander Bray who represented the international body we agreed conditions were suitable but we gave about ten minutes for the water to settle down.

As far as could be seen the Loch was perfect but I did pass a message which have subsequently heard was definitely delivered to John Cobb to the effect that conditions to the South were better than to the North.

After a few minutes I heard on the radio that the CRUSADER was off and shortly thereafter she appeared past Urquhart Castle.

I was about half way along the course alongside the road from Inverness to Fort Augustus.

As the CRUSADER approached at very high speed running beautifully I was very happy until I and the other spectators with me saw to our consternation a series of three "swells" or washes coming out into the Loch into his path.

There was nothing we could do at this stage.

He hit them with a smack at a speed of at least 240 - 250 m.p.h. and throttled back momentarily but then accelerated again with his engine.

However, although he seemed to steady himself momentarily, the boat was not planing steadily from that time onwards and built up an ever larger instability or "porpoise" ending in a nose dive into the water.

Speed by this time had been considerably reduced as he had again cut the engine presumably finding the boat uncontrollable.

From the fact that he was really in trouble from before half way down the mile and that speed from the time of hitting the swell was considerably reduced one must come to the conclusion the speed was in the neighbourhood of 250 m.p.h. when she hit the first swell.

The average for the whole mile was officially timed to be 206.8 m.p.h.

I am unable to form any real opinion as to what caused the instability after the first smack in to the system of "swells" reflected back from the shore.

Peter Du Cane's first, and previously unpublished, statement differs from his second statement that was released later, and was said to have been written "within days" of the accident. *Hampshire Archive/Du Cane archive*

I was about half way along the course alongside the road from Inverness to Fort Augustus.

As the CRUSADER approached at very high speed running beautifully I was very happy until I and the other spectators with me saw to our consternation a series of three "swells" or washes coming out into the Loch into his path.

There was nothing we could do at this stage.

He hit them with a smack at a speed of at least 240 – 250 m.p.h. and throttled back momentarily but then accelerated again with his engine.

However, although he seemed to steady himself momentarily, the boat was not planing steadily from that time onwards and built up an ever larger instability or "porpoise" ending in a nose dive into the water.

Speed by this time had been considerably reduced as he had again cut the engine presumably finding the boat uncontrollable.

From the fact that he was really in trouble from before half way down the mile and that speed from the time of hitting the swell was considerably reduced one must come to the conclusion the speed was in the neighbourhood of 250 m.p.h. when she hit the first swell.

The average for the whole mile was officially timed to be 206.8 m.p.h.

I am unable to form any real opinion as to what caused the instability after the first smack into the system of "swells" reflected back from the shore.[256]

This original draft was written while Du Cane was still at Loch Ness, whereas the 'official', released statement, although claimed to have been compiled within hours of the tragedy, was actually written some time after. It makes only a passing reference to "modifications", without stating quite what these were, and indicates that Du Cane thought the hull was safe to *above* 200mph. Neither version of the report makes reference to the *Maureen* leaving her station, and suggests, contrary to the observations of others, that *Crusader* was "running beautifully", although this may be because Du Cane's vantage point was quite high. Neither statement mentions that Cobb did not like the larger rudder, which he felt was "constantly kicking back". This is not to suggest any subterfuge, but witnesses with incomplete memories of events can subconsciously 'fill in the gaps', especially if prompted.

If one takes the 'consensus of opinion' from the various statements, then we have to concur with Bert Denly's observations. Although he had only a

WHY?

rear view as *Crusader* sped away from him, and seen through binoculars, Denly was an experienced high-speed driver and in his position of 'second in command' to George Eyston had observed many record attempts, on both land and water.

> Having taken John out to *Astrid*, I stayed onboard as I could see a long way up the Loch. I knew the idea was to make a high-speed run, but I got the impression John, having seen the amount of people who had turned up to see a 'trial', was feeling pressure to give a bit of a show. There really were quite big crowds everywhere. Still John wasn't new to all this and he had agreed with Reid that he would wait the two or three weeks until he [Railton] got back.
>
> I kept out of the way as the batteries were plugged in and the engine started up — it was all very different from the old Rolls' R-type days in America. John throttled up, and started to move away, and was up out of the water in no time. It took me a while to pick him up in my sights as the spray cleared, but you could see him getting going, but the boat wasn't happy once he cleared the castle headland, the spray coming from the front was constantly changing, not staying the same as some other runs. It took a while for the sound to get to us, and I could see the bow bouncing, and as I heard the engine slow, she bucked, and disappeared in a cloud of spray. I heard a woman scream, then silence.[257]

John 'Lofty' Bennetts, the De Havilland engine technician, gave this account.

> [I recall that day] like it was yesterday. Mr Cobb was a giant of a man, really very largely built, but he was so quietly spoken, and did nothing but thank everybody for their help. Once we had the engine in and running, Mr Du Cane thought there might be a little too much throttle available, he told us the boat had a tendency to hold on to her speed on a closed throttle, and asked us if we could install a throttle stop, which we could on the hand throttle, but not on the foot throttle, but it became apparent that every pound of thrust was needed to get the boat up and away. Anyway, on the last day I was at the end of the pier with George [Bristow] who was in the boat ready to go out to the *Astrid* for the start-up, when Mr Cobb appeared next to me with

CRUSADER

2 Holmfirth Flats,
Holmfirth Road, Sea Point,
Cape Town.

8th Oct. '52.

Dear Commander Du Cane,

I had already left Aberdeen before your wire arrived, and it has taken some time to catch up with me – thus the delay in replying. Furthermore, as I am leaving London first thing to-morrow morning for Southampton and Rome, this must unfortunately be a very brief and hurried report, or eye-witness account of the accident on the 29th September.

At the outset, I should like to point out that although I served for 6 years as a Pilot in the R.A.F. during the war, and learnt quite a lot about thrust, drag, pressure waves, resistance etc., as applied to air travel, I am somewhat ignorant of these same factors at high speed through water. This report, therefore, is purely _factual_, and I shall not attempt to advance any _theories_ as to what happened during that fatal run, except to dispute one theory which I see has already been advanced, namely – that the three waves which the Crusader hit, were pressure waves set up by the Crusader itself.

Just before the start of Mr. Cobb's last run, I was standing in front of your car overlooking the measured mile, and at a spot, which I think you will agree, was rather nearer the South marker.

I was studying the surface of the Loch, which appeared from this position, to be absolutely ideal – glassy and no ripple. Just when the second verey-light was fired, however, I noticed a slow ripple emerge from the near bank, which I took to be a back-wash from the bow-wave set up earlier by the Maureen (I think that was the name of the boat which took the time-keepers out). When I noticed this, I expressed the hope that Mr. Cobb would keep well over towards the East side of the loch.

The next moment he appeared, rather close to the West bank. It was a thrilling sight, and the Crusader appeared to be going splendidly. Just as it passed me, I concentrated all my attention on the floats, as I was particularly interested to see what happened when they hit these three slow, shallow swells.

There were three distinct bumps, and although the Crusader's stern remained in the water, the front bounced rather viciously. I don't think the Crusader started breaking up at this stage, and as far as I am concerned, the rest is a matter of conjecture. I rather think that John Cobb reacted to these vicious bumps in the same way as the average motorist would react to speed-wobble, when driving a car – he throttled back. The Crusader then appeared to nose-dive into the loch before breaking up.

I trust that this report is satisfactory, and in conclusion I should like to say how sorry I am, that such a gallant British venture had to end so tragically.

Yours sincerely,

(Sd.) D. Tyndall.

Mr D. Tyndall's statement described the three waves as "a slow ripple" and "shallow swells". *Tyndall family archive*

WHY?

Mr Menzies. "Well, what do you think, 'Lofty?'" he said, and I said it looked better out on the loch than it had. "Yes, it will have to do, I shall see you later." And with that he climbed down into the boat with George [Bristow] and Bert Denly, and I watched him go out with Mr Menzies and Mrs Cobb.[258]

Even cursory study of the newsreel footage clearly shows an unhappy boat. Despite that, two of the witnesses, Peter Du Cane and Joyce Cronk, who were standing next to each other, stated that *Crusader* was running "beautifully" — they used the same word. This is at odds with what so many others said, although it should be noted that Du Cane and Mrs Cronk were at a higher vantage point than all the other observers quoted here. All the same, Du Cane's and Mrs Cronk's accounts differ in that the latter says she could see little through the spray whereas the former describes it quite clearly. Only these two accounts and one other, from Mr. D. Tyndall, allude to any wave formations, and Tyndall's describes them as merely a "slow ripple". His statement read:

> At the outset, I should like to point out that although I served for 6 years as a Pilot in the R.A.F. during the war, and learnt quite a lot about thrust, drag, pressure waves, resistance etc, as applied to air travel, I am somewhat ignorant of these same factors at high speed through water. This report, therefore, is purely <u>factual</u>, and I shall not attempt to advance any <u>theories</u> as to what happened during that fatal run, except to dispute one theory which I see has already been advanced, namely — that the three waves which the *Crusader* hit, were pressure waves set up by *Crusader* herself.
>
> Just before the start of Mr. Cobb's last run, I was standing in front of your car overlooking the measured mile, and at a spot, which I think you will agree, was rather nearer the South marker.
>
> I was studying the surface of the loch, which appeared from this position, to be absolutely ideal — glassy and no ripple. Just when the second verey-light was fired, however, I noticed a slow ripple emerge from the near bank, which I took to be back-wash from the bow wave set up earlier by the *Maureen* (I think that was the name of the boat which took the time-keepers out). When I noticed this, I expressed the hope that Mr. Cobb would keep well over towards the East side of the loch.

The next moment he appeared, rather close to the West bank. It was a thrilling sight, and the *Crusader* appeared to be going splendidly. Just as it passed me, I concentrated all my attention on the floats, as I was particularly interested to see what happened when they hit these three slow, shallow swells.

There were three distinct bumps, and although the *Crusader*'s stern remained in the water, the front bounced rather viciously. I don't think the *Crusader* started breaking up at this stage, and as far as I am concerned, the rest is a matter of conjecture. I rather think that John Cobb reacted to these vicious bumps in the same way as the average motorist would react to speed-wobble, when driving a car — he throttled back. The *Crusader* then appeared to nose-dive into the loch before breaking up.[259]

It could be that in the weeks after the accident but before his statement, Du Cane had had the opportunity to see film footage, but it does appear that by discussing and sharing recollections, his and Mrs Cronk's statements are a little compromised. It is curious that there are so many similarities in their statements that do not match the observations of people who we know were closer to the event.

One inconsistency concerns Mrs Cronk's whereabouts at the time of the accident. Many years later, Vicki Cobb stated that, on hearing about the accident over the public address, which was only available in the area around the project base at Urquhart Bay, it was Joyce Cronk who accompanied her to the privacy of the caravan. Arthur Bray also confirmed that Mrs Cronk was present when he broke the news of John Cobb's death to Vicki, again in the caravan.

Yet at the time of the accident she was supposed to have been with Du Cane and another Vosper man in their radio car, parked near the halfway point of the measured mile course, as confirmed by Vosper's own report map reproduced here. This was a 5.3-mile drive from the project base and took about 17 minutes to reach on clear roads. Du Cane's account stated that they only arrived back at Urquhart Bay after the search and recovery, which he said he watched through binoculars from the roadside to ascertain Cobb's condition before he set off, arriving back as Cobb's body was being brought ashore.

There are no doubts about where Mrs Cobb was. Hugh Patience, Alec Menzies, Bert Denly and George Eyston were all quite clear that she stepped out of the caravan after her husband left, walked to the shore, and watched

WHY?

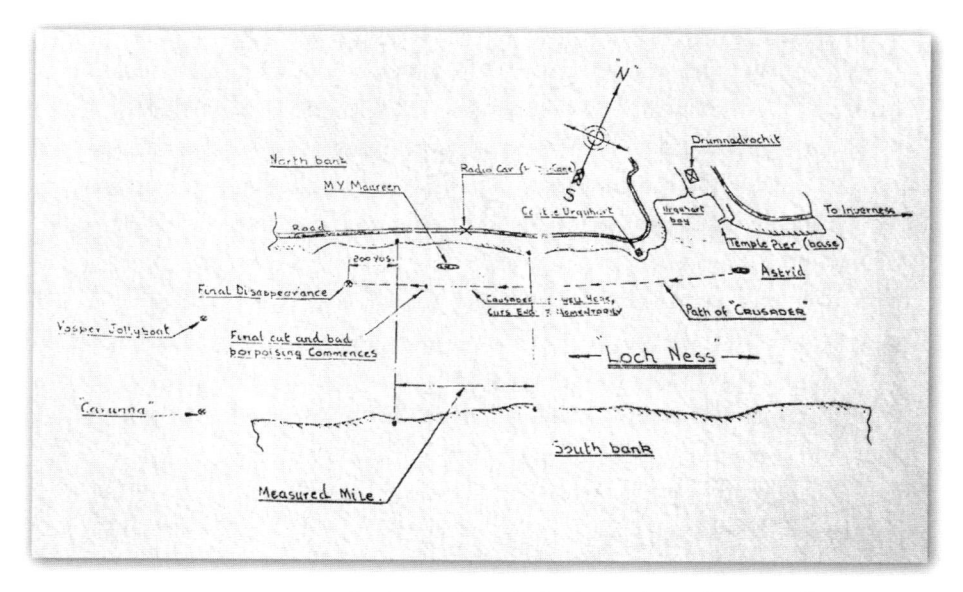

The map produced by Vosper for its crash analysis shows the radio car on the northern shore annotated "Du Cane". Hampshire Archive/Du Cane archive

from there. Indeed, Menzies wrote a magazine article in which he stated: "A fast launch took him [Cobb] out to CRUSADER and I and Mrs Cobb, with several others, watched from the end of the pier. The engine started and in a few moments CRUSADER was on her way. Skimming the water like a bird she moved beautifully, John Cobb giving her full throttle as she passed close by Urquhart Castle. In what seemed but a few seconds we heard the engine cut out."[260]

Ted Cope, the PYE radio engineer, recalled: "Mrs Cobb listened in to all the transmissions from my car by the shore, as it had a test radio installed. She said she liked it that way as she felt she wouldn't get in the way, and she thought my car was comfortable."[261]

From all this, one has to consider the possibility that in fact Peter Du Cane and Joyce Cronk were not, after all, at their claimed vantage point when the fatal run occurred and instead were watching from the same place as Mrs Cobb by the shore. Despite their statements, perhaps Du Cane had felt that it was not possible to drive back out to the radio car position in the time available because the road was heavily congested. The *Glasgow Herald*'s report of 30th September not only stated that Du Cane "was standing beside Mrs Cobb when the accident happened" but also that "thousands of sightseers" were there to watch.

CRUSADER

A statement from one witness is informative about Du Cane's location at this time. Mr McNaughton was foreman at MacRae & Dick, the Lucas dealer in Drumnadrochit, and was present as his team was responsible for charging the batteries for the radio systems. There were three full sets of batteries: as one was being used, another was held in reserve, and the third was charging. McNaughton had taken fresh batteries to the project base himself that day from his home, having been warned of a possible early start. "I had unloaded the batteries, and when I turned to leave I saw there was very little chance of getting back in should I leave, so I decided to stay put, as I had the reserve set in the company van. So I watched as the boats and the radio cars arrived back, and got invited to breakfast when Mr Cobb returned and put a hold on. A car had brought back some of the Vosper people, one of the radio cars, and I asked the driver about the state of the roads and he said that he and Mr

Vicki Cobb sits in Ted Cope's car listening to radio messages from *Crusader*. *Getty Images/Ray Kleboe*

WHY?

Du Cane would probably have to stay put now as the roads were more or less blocked."[262]

Various observations about "the state of the roads" beg the question why there were so many spectators present. No record attempt was being made, and no one had issued a statement to the effect that it was, but movement around the loch was almost impossible that day. It has also been suggested that the crowds were swollen with schoolchildren who had been given that Monday morning off school to watch proceedings, but Gordon Menzies, Alec's son, is quite clear that this was not the case. So, what had drawn 'thousands' to watch a mere test run?

Joyce Cronk's statement is the only one that says there was an explosion. Although Arthur Bray and Frank Lydall were much closer, they did not comment about any loud event.

Then we have the memories of the

> Cobb's achievement.
> "We have lost a very gallant Briton who has proved that as a nation we can go out and do things," he said. "What he did was to be the first man to travel on water at over 200 miles per hour."
> Three big waves on the second part of the measured mile may have caused Mr Cobb's speedboat to nose-dive into the loch. This is the theory advanced by Mr Peter du Cane, managing director of Vospers, Ltd., the builders of the craft. Mr du Cane, who was standing beside Mrs Cobb when the accident happened, was an eye-witness. Mr du Cane said:—"He was a very great sportsman. He must have been travelling at 240 m.p.h. to average 206.8 m.p.h. on the measured mile. He hit three big waves—that was his trouble."
>
> **Seemed to Somersault**
> It was apparent to the thousands of sightseers that Mr Cobb's speed was well in excess of the 178.4 m.p.h. American-held record. Just before reaching the second time-keeper's hut at the end of the measured mile, however, the boat developed an alarming bounce, which became rapidly more pronounced. In a flash it seemed to somersault and in a cloud of smoke

The *Glasgow Herald*'s report stated that Peter Du Cane was standing beside Vicki Cobb at the time of the accident.
Glasgow Herald archive

unofficial 'official' photographer. Ray Kleboe created a unique place for himself in the annals of post-war magazine publishing. Returning from six years of war, he took up an offer from Stefan Lorant, founding editor of *Picture Post*, the leading national weekly magazine of the time. As a freelance for *Picture Post*, Kleboe was always considered something of a 'loner' but, by working alone, he got into places no other photographers ever did. Commissioned by Castrol's Laurie Sultan, he photographed the *Crusader* project almost from the day the first piece of wood was cut, and by September 1952 he was at Loch Ness. Nearly all of the most recognised pictures of *Crusader* — and of John Cobb and Reid Railton at that time — are Kleboe's work. Although he died in 2005, aged 91, his recollections of his time with the project were vivid and detailed:

CRUSADER

The A82 on 29th September, at the point where the Vosper radio car was normally stationed. Despite being somewhat indistinct, this photo confirms that the road was probably impassable. *National Library of Scotland*

I arrived more or less the day John and Reid arrived, and there was a huge sense of excitement and anticipation about the place. John saw to it I could be wherever I thought I should be to capture the event, and the scenery and light added enormously to the possibilities.

But there was apprehension as well, and I was asked not to take pictures of the engineers as they examined the broken screws they seemed to take out on a daily basis, and there was an unease about John's wife that was so obvious to the lens as she watched the series of test runs made in the recent weeks. We watched from the end of the pier that final day, and I couldn't, simply couldn't photograph her as she put her face in her hands as the boat disintegrated. Friends led her away from the pier in tears even before his death was confirmed.[263]

And so we have many statements on record that simply do not agree as to who was where at the time of the accident.

At the time of the crash, Reid Railton was still on his way to Southampton and, on receiving a message about the tragedy from George Eyston, immediately returned to Loch Ness, arriving just as Peter Du Cane was about to depart. Bert Denly recalled Railton being "ashen-faced" as he spoke with Du Cane and noticed that the two "exchanged some papers".[264] By the evening, Railton had seen the need to gather evidence and set about arranging survey maps, access to all available film images and any statements that had been made.

No official investigation into the accident took place, although the press made much of discussions between Du Cane, Railton, Cobb and Eyston regarding the front planing shoe, as well as the three waves.

WHY?

Crusader's tendency to hold her speed (and sometimes even accelerate) on a closed or partial throttle setting had worried both Railton and Du Cane but was not commonly known. The BBC film of the accident is by far the closest and best-quality record of the event, and, viewing it, one could be excused for believing that the front planing shoe did not, as some thought, collapse — or, if it did collapse, perhaps it did so only partially — but again few were party to the boat's long gestation and construction methods. What is clear from all four films is that something caused *Crusader*'s fore/aft stability to become severely compromised — and this occurred very early in the run.

Another factor, often ignored, was Cobb's role in that fateful last run. The suggestion put forward by some, including Peter Du Cane, was that he was under pressure to get the record as quickly as he could, because of business commitments, whereas others, such as George Eyston, believed that he just wanted to put on a show. For Cobb, speed was his respite from business, and something he would not give up. It is true that he had just been appointed Chairman of the Falkland Islands Company and it has been claimed that he was due to depart for a three-month trip to the Falklands, where the company owned 1.2 million acres of land and some 300,000 sheep. There was also supposedly an agreement that Cobb would no longer involve himself in risky high-speed ventures after taking up his new position, adding to the pressure on him to succeed at Loch Ness. Neither of these possibilities seems to have substance.

Cobb more or less owned the Falkland Islands Company and his partners were under no illusion that his speed exploits were, as he put it, his "escape from the mundane" and as such not negotiable, and he had stated in a letter to Railton early in 1951 that he had no deadline by which to set a record, "not this year or next my friend".

His appointment as Chairman of the Falklands Islands Company had come about only two weeks before his death, so it seems unlikely that a visit to the Falklands — especially a prolonged one — could or would have been organised in that short time. In any case, it does not seem plausible that Cobb, even if invited into a directorial position, would have felt the need to take a three-month trip to look at some sheep, at a time when his dealings were mainly in more valuable mink and sable. Furthermore, Cobb's personal paperwork shows that his appointment also encompassed the company's shipping operations in the UK. In fact, Cobb had been involved with the Falkland Islands Company for some years beforehand, following in the footsteps of forebears from the previous century, so there would have been no need for him to rush off on a trip to deal with something that was already part of his day-to-day work. Nor would it have been his first trip: he had visited the Falklands in 1946, when he flew from the newly opened London Airport (later renamed Heathrow) to Montevideo, Uruguay, and sailed to the islands on the first voyage of the SS *Lafonia* after its name change from SS *Perth*.

Where Cobb may have felt under pressure was from his new-found 'friends' in the Scottish Highlands.

Bert Denly recalled many years later: "There's no doubt in my mind that he wanted to put on a show that day, maybe not the record, but he felt he owed the locals something for their patience, and he was very taken aback by quite how many had turned out, and who had all waited since very early that morning. George had said, from the evening before, and that morning, it was a high-speed trial, not an all-out attempt, but we hadn't had the team briefing that Railton insisted on, and John must have been aware as he and George were always conferring."[265]

The first part of Denly's observations were eerily echoed more recently by Doug Nye, who, as quoted in Chapter 1, said: "Like many shy or enclosed people, he evidently quite liked the opportunity to perform before a large crowd, as long as he didn't have to address them or look them squarely in the eye."[266] Running on Loch Ness on 29th September 1952 certainly gave Cobb that opportunity.

Eyston, in an interview with Raymond Glendenning of the BBC, later implied that the team had found the structure of the boat inadequate yet had let her run, because of "outside pressures".[267] But again, the activity of 29th September was not a record attempt, so the attendant crowd and pressures seem out of place.

However, we then have a letter written by Peter Du Cane on 6th June 1970 in which he wrote: "I would probably change the planing surfaces to avoid the rough ride but not much else for the next year when I anticipated 300 MPH", adding, "We had actually gone so far as to design and draw out the required vee'd planing surfaces."[268]

Not only does this beg the question as to why Du Cane resisted the use of vee'd surfaces while all the time planning to use them, but it also directly contradicts his previous statements that Cobb went against his advice to run on the fatal attempt because he was in a rush to get the record due to outside

WHY?

A still from the Movietone News footage shows that there were actually many wakes left on the loch, but the biggest, and 'first of the three', is running at the wrong angle to have been created by the *Maureen*. British Movietone

influences, or an agreement to his company that would mean 1952 would be his last record attempt. Vosper was already working on a 1953 attempt.

From studying all the statements, film clips and reports, the simple facts are that *Crusader* left her berth alongside the *Astrid* and accelerated briskly, entering the measured mile at approximately 190/200mph and continuing up to 230mph at the half-mile point of the measured mile. During the run-up, a progressive vertical porpoise set up at the bow, visible as a turbulent wash that could be seen travelling fore and aft along the line of the forward planing 'wedge', from bow to step and back, as that part of the hull at first rose, planed, then dropped again, although not enough to drop off the plane.

About halfway through the measured mile, *Crusader* encountered a change to the surface of the water, seemingly in the form of three waves. Films and photographs confirm the presence of these waves and over the passing years there has been much discussion as to their source, the usual conclusion being that they were 'reflected' from the shore. Alec Menzies's son Gordon, who was present at the attempt as a schoolboy, and who still operates from Temple Pier, is adamant that these waves were not reflected in that way:

> I do not dispute that this happens in bodies of water such as canals but a careful look at Loch Ness and its shores — at water level — shows

that the Loch's shores are not 'steep' but are almost horizontal for a width of one metre or more and do not consist of solid rock faces. The shores in the area of the run consist mainly of rounded cobbles of varying sizes which absorb almost all of the power of any waves hitting them. Any reflected portions are of low strength and a fraction of the height of the primary waves which caused them.

Personally I believe that this [reflected wake theory] was a sop created to salve consciences feeling guilty for not waiting long enough to let the disturbances caused by the attendant craft to die down but also for not recognising that the design of the forward shoe was fatally flawed.

Regardless of their origin, ridges of water did cross the course, moving obliquely to *Crusader*'s direction of travel. On meeting the first of these waves, a loud "smack" or "crack" was heard by observers, and the throttle was also heard to reduce. After the first wave, the throttle was heard to be "re-applied", then as the craft encountered the second undulation, and bucked again, the throttle stayed at a constant level, as if the craft was being "driven" through the problem. The bow barely deflected and, as designed, started to drive back down into the water to regain stability. Then, before the bow hit the third wave, *Crusader*'s natural hydrodynamic lift brought the bow back up, considerably above its normal ride height. It is then very evident from the BBC film that the bow was high enough to merely skim the crest of this last wave, but the rear sponsons, travelling at the correct operational height, made contact. Their large, flat surfaces caused the sponsons to climb up the wave and tip the bow into the water in what Du Cane later described as "a kangaroo leap in comparison to the previous hops".[269]

As marine architect Lorne Campbell observed: "In powerboat racing this is called 'tripping', where the aft of the hull catches a wave which kicks the stern up and rotates the whole boat nose down. In *Crusader*'s case the bow-down moment caused by the jet thrust would have contributed considerably."[270]

Furthermore, it is very unlikely that there were only three waves, as such undulations cannot appear in isolation. It is more likely that the feature was created by the interaction of the wakes of all the craft on the loch at the time, as wake waves appear as a continuous line.

Little data from the run was available at the time — merely the film and witness statements. Although Cobb's land speed car had been fitted with a basic form of data-logging, made by Greecen, it was found impossible to

WHY?

include the system on a boat hull because of the vibrations, and in any case there was little point because measurable parameters were few. As such, the hull contained only an accelerometer, on loan from the Marine Aircraft Experimental Establishment at Felixstowe. The data recorded by this device, which was beyond recovery, could only have been the craft's acceleration and peak speed.

As the months passed after the accident, no official enquiry was deemed necessary, although both Peter Du Cane and Reid Railton continued to consider what had gone wrong.

Both Railton and Arthur Bray were sceptical about the three waves. Indeed, feelings between Bray and Du Cane on the matter were so strong that the two would not speak for some years, Bray adamant that any such waves, even if they existed, were not the cause of the crash.

Many years later, Bert Denly recalled that, for him, first mention of this wave formation occurred the day after the accident, when many of those involved in the project — including Basil Cronk and Du Cane — took a trip out on the loch. On the return journey to shore, Cronk mentioned in passing that he could make out the wake from their outbound run reflecting from the shore. Whether or not this conversation took place before Du Cane wrote his first statement cannot be known.

On the matter of Loch Ness's conditions and behaviour, there were only two local men — Hugh Patience and Alec Menzies — whose views John Cobb considered important. Patience, owner of the *Astrid*, remarked that these 'wakes', when they appeared, were without any force or density, and would have had no bearing on the outcome, a view that Menzies supported.

Although some people refute that wakes on Loch Ness reflect, there is the possibility that the topography of the loch can create a form of Soliton Effect under water that may have damaged *Crusader*. The Soliton Effect, first described in 1834 by John Scott Russell, is the trait for the bow wave of a speeding vessel in a narrow stretch of water to become detached and continue separately ahead of the hull. It has to be said that Loch Ness, although not of great width compared with its length, is not comparable with a narrow canal, where the Soliton Effect can be quite commonly observed with displacement craft, but the loch's underwater topography, relative narrowness and the movement of various tender craft mean that it should not necessarily be ruled out.

CRUSADER

Loch Ness expert Adrian Shine had this to say on the matter:

> I would not use the word Soliton. The loch, even inshore, is very deep and, of course, one side of any parallel course is unconstrained.
>
> I simply believe that the run commenced before the boat wakes had dispersed. My understanding is that the *Maureen* had just dropped the timekeepers. A series of waves from a vessel which stops at any point will continue forwards as well as the usual displacement wakes from this and other vessels. I see no need to consider underwater phenomena. I just don't think Cobb's team were aware of the persistence of wakes in general, despite their previous tests. The explanation they came up with the next day, about wakes reflecting from the shore, is completely incorrect, as many of our experiments have demonstrated.[271]

Interestingly, on 30th September Arthur Bray and Peter Du Cane were seen having a heated discussion amidst the upsetting task of clearing Temple Pier. Bert Denly: "There was a definite atmosphere between them, both waving wads of paper at each other, waving and pointing out to the loch. I wasn't eavesdropping, but you didn't have to, it wasn't a quiet chat they were having. But I do remember Mr Bray saying, 'That is not what I saw.'"[272]

The next day, 1st October, Bray wrote his official observer's report to the Marine Motoring Association. It contradicts his 'on the spot' report made within hours of the accident, but certain parts raise other interesting points.

MR. JOHN COBB'S RECORD ATTEMPT ON LOCH NESS

In connection with the above, I proceeded to Inverness, arriving A.M. September 18th, as the appointed official observer.

I was then taken to Temple Pier, Drumnadrochit and with the official timekeepers, the surveyor, and other officials I made an inspection of the course by water, including the sighting of the statute mile posts on the north-west side of the Loch. At each end the transit posts were properly constructed from concrete and were sighted in transit with four similar posts on the other side of the Loch. I satisfied myself from explanations given by the surveyor with the chart that the course had been laid out to comply with the U.I.M. rules and was accurate.

I consider that both this course and others which could be laid out,

WHY?

make Loch Ness the best venue in the country for high speed record attempts and also racing, if more accommodation can be provided for launching, hauling-out and servicing craft. This view is in no way changed as a result of this tragic accident.

On September 19th, when conditions became favourable, it was possible to run CRUSADER for two runs over the mile before 08.30, with the following results:-

N.E. to S.W. run 185.567 statute m.p.h.
S.W. to N.E. run 160.714 statute m.p.h.
Mean speed 173.140.

The reduced speed of the second run was caused by difficulty in CRUSADER getting on her step after turning.

The run was made about ¾ mile out from the timekeepers' posts on the north-west side, which was considered too far for accurate sighting, and, as a result, I requested Mr. Cobb to make further runs within, if possible, ¼ mile from the timekeepers, which he did on his last fatal run. The Loch on this occasion had not a mirror calm and the water was slightly disturbed, but in spite of this CRUSADER planed perfectly on each run without any fore and aft motion, but was a little difficult to control in the side gusts with the small rudder.

My observations were made from a 60 ton yacht MAUREEN IV, kindly placed at our disposal by Mr. M.D. Hodge, and on this I took up my usual position about half a mile along the mile and about 500 yards off shore, after landing the timekeepers at their posts by dinghy. The PYE Inter-Com communications, which had been arranged, were excellent, and consisted of two Dolphin sets in each timekeepers' hut on shore and one in MAUREEN, this provided secret communication between the two timekeepers and myself, which was helpful in obtaining times and speeds immediately. In addition, PYE V.H.F portable sets were provided in MAUREEN, the box at the pier, the drifter depot ship, two speed boats, an automobile and CRUSADER. It was thus possible to obtain reports on surface conditions at each end of the mile from the speed boat and from the car, which took up position on high ground half way along the mile looking down on the Loch. In addition, it was possible to give warning of the necessity of postponing any run by direct communication to the boat shed, according to weather conditions. It is, however, suggested that in future, reports on weather conditions, etc. should be made by official observers rather than parties interested in commerce, and that the official decision regarding the start or delaying of an attempt must be made by the observer, after he has satisfied himself that the timekeepers are ready and that the course is free from driftwood, etc.

In connection with the latter, the organisation for spotting and collecting driftwood was excellent, and consisted of periodical inspections by the speed boats running over the mile, and by my personal observations from MAUREEN, I am quite satisfied that at no time during the preliminary runs or during the fatal run was there any driftwood on the course.

After waiting from the 19th until the 27th September for the weather to moderate, and after the rudder had been changed on CRUSADER another run was made at 08.50 on the 27th September, with the following figures: -

N.E. to S.W. speed 152 statute m.p.h.

S.W. to N.E. speed 100 statute m.p.h.

Mean speed 116 m.p.h.

This run was more of a test for the turning qualities of the boat rather than speed, as it was then found possible for Mr. Cobb to turn CRUSADER through 360° at each end of the course without outside assistance, and without coming off the step.

Bad climatic conditions continued until the 29th September when we stood by from 07.50 until 09.50, but although conditions were better than on the other occasions the Loch was not mirror calm. The personnel were then stood down, at Mr. Cobb's request, until approximately 11.20, and both Commander du Cane in the car on the high ground and myself agreed that conditions were favourable for the attempt, as the Loch was mirror calm but with one or two small <u>ripples</u> on either side of the course.

This information was conveyed to Mr. Cobb at the drifter, I fired two Vereys 'All Clear' lights, and he then immediately proceeded with CRUSADER over the mile at approximately 11.55.

On his north-east to south-west run he came into the mile at a speed in excess of his previous runs, and CRUSADER was planing beautifully on a steady course without any 'porpoising', and was fully under control. Immediately he passed me on the ½ mile

WHY?

OFFICIAL OBSERVER'S REPORT

To The President of the Marine Motoring Association Ltd.

THE LATE MR. JOHN COBB'S RECORD ATTEMPT ON LOCH NESS.

In connection with the above, I proceeded to Inverness, arriving A.M. September 18th, as the appointed official observer.

I was then taken to Temple Pier, Drumnadrochit, and with the official timekeepers, the surveyor, and other officials I made an inspection of the course by water, including the sighting of the statue mile posts on the north-west side of the Loch. At each end the transit posts were properly constructed from concrete and were sighted in transit with four similar posts on the other side of the Loch. I satisfied myself from explanations given by the surveyor with the chart that the course had been laid out to comply with the U.I.M. rules and was accurate.

I consider that both this course and others which could be laid out, make Loch Ness the best venue in the country for high speed record attempts and also racing, if more accommodation can be provided for launching, hauling-out and servicing craft. This view is in no way changed as a result of this tragic accident.

On September 19th, when conditions became favourable, it was possible to run CRUSADER for two runs over the mile before 08.30, with the following results :-

        N.E. to S.W. run 185.567 statute m.p.h.
        S.W. to N.E. run 160.714 statute m.p.h.
        Mean speed 173.140.

The reduced speed of the second run was caused by difficulty in CRUSADER getting on her step after turning.

The run was made about ¾ mile out from the timekeepers' posts on the north-west side, which was considered too far for accurate sighting, and, as a result, I requested Mr. Cobb to make further runs within, if possible, ¼ mile from the timekeepers, which he did on his last fatal run. The Loch on this occasion had not a mirror calm and the water was slightly disturbed, but in spite of this CRUSADER planed perfectly on each run without any fore and aft motion, but was a little difficult to control in the side gusts with the small rudder.

My observations were made from a 60 ton yacht MAUREEN IV, kindly placed at our disposal by Mr. M. D. Hodge, and on this I took up my usual position about about half a mile along the mile and about 300 yards off shore, after landing the timekeepers at their posts by dinghy. The Pye Inter-Com communications, which had been arranged, were excellent, and consisted of two Dolphin sets in each timekeepers' huts on shore and one in MAUREEN. This provided secret communication between the two timekeepers and myself, which was helpful in obtaining times and speeds immediately. In addition, Pye "V.H.F." Portable sets were provided in MAUREEN, the box at the pier, the drifter depot ship, two speed boats, an automobile and CRUSADER. It was thus possible to obtain reports on surface conditions at each end of the mile from the speed boats, and from the car, which took up position on high ground half way along the mile looking down on the Loch. In addition it was possible to give warning of the necessity of postponing any run by direct communication to the boat shed, according to weather conditions. It is, however, suggested that in future, reports on weather conditions, etc. should be made by official observers rather than parties interested in commerce, and that the official decision regarding the start or delaying of an attempt must be made by the observer, after he has satisfied himself that the timekeepers are ready and that the course is free from driftwood, etc.

First page of Commander Arthur Bray's official report. This differed from his original — and commented on "outside commercial interests" being brought to bear for *Crusader* to run on 29th September. *Marine Motoring Association/FIM archive* mark at a distance of approximately 200 yards from MAUREEN, CRUSADER developed a <u>very slight</u> 'porpoise', far less than I have seen similar boats 'porpoise', and certainly one which I would describe as being fully under control. Immediately following this CRUSADER developed another slight 'porpoise' and the engine was slightly reduced. CRUSADER then developed a third 'porpoise' slightly worse than the first two and the engine then obtained again its maximum revolutions. Within a fraction of a second after this, and after CRUSADER had passed the south-west mark, she dived slightly to port in a fraction of a second and completely disappeared. There was no explosion but a certain amount of steam, due no doubt to the contact of the hot engine with the cold water. A small amount of debris, which was thrown into the air to approximately 20 ft, was, no doubt, caused by disintegration of the hull by water pressure in the dive. The approximate time was 11.56.

I then proceeded in MAUREEN at full speed to the centre of the wreckage in approximately 7 minutes, where I stopped and found no trace of anything except wreckage within a radius of about 30 yards. I then spotted Mr. Cobb's life jacket about 20 yards outside the area of the wreckage, and, at the same time, the Vosper 'Jolly boat', which had been coming up at full speed from the south-west end of the Loch arrived, and was able to take Mr. Cobb on board. Although at that time I was not certain, I formed the opinion that Mr. Cobb was dead. I instructed the 'Jolly boat' to come alongside MAUREEN, and, in the meantime, I telephoned to the base for an ambulance and requested that the pier be cleared, and that Mrs. Cobb was not present when we returned.

In the meantime, the deck saloon of MAUREEN had been prepared for the eventuality of bringing Mr. Cobb on board for first aid. Immediately the 'Jolly boat' came alongside I descended and found Mr. Cobb lying on top of the engine case with his clothes and life jacket intact. The lower part of his jaw was badly torn and his right leg was useless. His pulse was negative. I then instructed the 'Jolly boat' with me to proceed at full speed to the pier. When away from MAUREEN and photographers I stripped back some of his clothes and after listening to his heart I confirmed that he was dead. Half-way along the course to the pier the doctor arrived in another speed boat, and after removing further clothing confirmed my fears. His body was then landed on the pier awaiting the ambulance (which did not arrive until 1 o'clock), and I then immediately gave instructions for all the speed boats and other craft to return and pick up as much wreckage as possible.

During the last fatal run I would emphasise that the Loch was virtually a mirror calm and that during this run I did not observe any wash or waves, but a very slight ripple, which has since been confirmed from the actual photographs and newsreels. I am confident that the water conditions as on previous runs were in every way favourable for the attempt to be made.

I subsequently confirmed personally to Mrs. Cobb, Mr. Cobb's tragic ending, and returned to Inverness, where in the afternoon I identified Mr. Cobb in the mortuary and made a verbal report to the Inspector of Police of the whole of the incident. The next afternoon, the doctor, after a post mortem, confirmed that death was instantaneous and that it was caused by 'shock through severe internal injuries'.

I would like to pay tribute to the efficiency of the timekeepers, Mr. Philip Mayne in the first runs, and to Mr Philip Mayne and Mr. R. L. Christian on the fatal run, who confirmed the official speed at 206.89 m.p.h. and to Messrs. Buckley, Lydall and Boots of the W.M.B.S.C. for their voluntary and generous assistance.

I should in addition like to speak highly of the services rendered by the crew of Mr. Hodge's MAUREEN and Mr. Hodge himself, and likewise of the efficiency of Mr. Harry Cole and Mr. Hugh Jones of the Vosper 'Jolly boat', who, in my presence, picked up Mr. Cobb's body within 8 minutes of the accident.

In conclusion, I would confirm that during the long and tedious waiting periods, John Cobb's usual calmness and patience prevailed, which was an outstanding example to all those connected with the organisation. The voluntary task he had set himself of regaining the water speed record for this country was typical of his tenacity of purpose, pluck and devotion to the cause of motor boating. My last tribute to his memory is that, his personal friends in the whole country have lost a very gallant gentleman who gave his life in an attempt to achieve his great ambition.[273]

We can only wonder why Bray changed the wording from his initial statement ("the bow was clearly bouncing in and out of the water well before the survey point for the mile") to use the phrase "planing beautifully", echoing the Vosper-related statements. What is evident is that he was plainly trying to avoid alluding to any wave formations other than normal water swells or ripples, and appeared to be angry that runs were being made at the suggestion of those with a commercial interest — which can only have been Vosper as Wakefields acquiesced to Eyston, who in turn relied upon Railton. This is a more likely reason for Bray and Du Cane refusing to talk to each other for years afterwards.

It is also notable that Bray makes no reference to the *Maureen* returning to Temple Pier, but does confirm that he knew Vicki Cobb was at the pier, and not in the radio car. He was also quite certain that it *was* a record attempt.

Regardless of this, from many of the statements it is obvious that *Crusader* was in trouble well before entering the timing area.

Peter Du Cane, for his part, would revisit his investigation into the *Crusader*'s end from time to time over the following years, but never really changed the views he stated in his first officially released statement. His initial analysis was of limited scope and restricted mainly to what could have been done differently. For his part, he seemed convinced about what he thought had gone wrong and never wavered from that view, although he did concede in later years that he and Vosper had made mistakes in the test and build phase.

As late as June 1970 Du Cane was still writing, "I would probably have changed the planing surfaces to avoid the rough ride", so why did he resist doing that at the time? What would have influenced that decision? Although, as we have seen, he envisaged that *Crusader* would run with vee'd planing surfaces in an attempt at 300mph in 1953, it is also apparent that he did not want others making statements that disagreed with his own — as is evident from various comments in his handwriting — and actively sought to discourage writers or researchers from speaking to some eyewitnesses until he had spoken to them first.

Reid Railton's analysis was overshadowed by the fact that John Cobb had been a long-time friend. It was some time the following year that he finally had it confirmed that the three-hoop armature had not been included in the final build, and concluded that *Crusader* had suffered a massive structural failure as a result.

Bill Maloney was involved in the initial study of *Crusader*'s accident and later, in 1967, he also investigated the crash that befell *Bluebird K7*. A highly

WHY?

qualified marine architect and engineer, his input came at the behest of the insurance companies covering the craft, but spilled over because of his work with Ken Norris on *Bluebird K7*. In 1958, Maloney collaborated with Douglas Phillips-Birt on an article for *Shipbuilding & Shipping Record* and included key extracts from his report:

> Working from the cine film, it is clear that *Crusader* was experiencing a violent vibration considerably exceeding 2.6 cycles per second.
>
> *Crusader*, travelling in excess of 200 mph ran in to a reflected system of waves, which appeared in her path at just about halfway through the measured mile. From the moment of impact with the first wave of this system, which was heard by many as a very audible 'smack', until her final nose dive in, *Crusader* gradually increased this longitudinal instability (of a resonant character, consisting of a combination of vertical oscillation and heave and pitch). More of an extreme 'flutter', as found in high-speed aircraft, than the more common 'porpoising' of a very high-speed boat, and it should be noted that this phenomenon was absent from the free-running model tests and from the previous runs of the full-size hull in speeds exceeding 200 MPH.
>
> George Eyston helped to trace the waves shortly after the accident and it is clear that an already vibrating hull had compounded this, and magnified after running through this wave system. They were referred to in statements from a number of onlookers. The cause of this wave train has not been exactly established. The bow wave train of a typical round-form craft leaves a most persistent pattern diverging at an angle of approximately 20 degrees its centreline, and in calm water it can last for a considerable time in the absence of a damping medium other than just the viscosity of the water itself.
>
> Where there are steep sides (such as is present at Loch Ness) the waves may be reflected at an angle of almost 60 degrees backwards and forwards across the Loch many times. Assuming the craft that created the waves had a speed of 12 knots and these waves would continue to spread at an angle of about 60 degrees to the course at a speed of 0.423 knots and would advance up the loch at approximately one fifth speed of the originating vessel.[274]

And there the analysis would have ended, had it not been for Donald Campbell.

CHAPTER 12
FROM THE ASHES

After Donald Campbell's failure to raise his father's water speed record in the Vosper-built *Blue Bird K4*, and its demise in 1951 when the gearbox tore loose from its mountings, he had decided that, as John Cobb was sure to take the record with *Crusader*, he would attempt to win the Harmsworth Trophy and return it to Britain.

Before *K4*'s final flourish, she had been converted to become a 'prop-rider', with the help of Reid Railton, Ewan Corlett and a young engineer called Lewis Norris. While Lewis was overseeing *Blue Bird*'s conversion, one of his older brothers, Ken, had become involved with the rival team of Frank and Stella Hanning-Lee. Ken had been apprenticed to the Armstrong Whitworth aircraft company at Whitley and by 1945, aged only 23, was in charge of the mechanical testing department. At the same time he also taught at Coventry Technical College, where he first grew his trademark moustache: "At one time I got up to give my lecture and all the students were laughing and nudging each other because they thought I was just one of them larking around, pretending. I must have looked younger than most of them. So I grew the moustache."[275]

While "brother Lew", as Ken always referred to him, was already working on *Blue Bird K4*, Ken enrolled at Imperial College, London to study aeronautical engineering, as well as taking business administration part-time at the London School of Economics. It was at this time that he met the Hanning-Lees and became involved in their *White Hawk* record project after being invited to "produce the stress figures for their 200mph water speed record boat". Ken was to have a rude awakening: "They only really had a full-size outline, chalked on a basement wall. I suddenly realised that I was going to have to do everything, and I was virtually imprisoned in this tiny room drawing away, until eventually it dawned on me that there wasn't going to be any payment."[276]

Lew Norris had been working with another older brother, Eric, at Kine Engineering, a company part-owned by the Meldrum brothers, who were relatives, and where Donald Campbell was managing director. Once Ken had made his escape from the Hanning-Lees, the three brothers joined forces. On New Year's Eve of 1952, shortly after Ken, Eric and Lew had set up Norris Bros, an engineering consultancy in Burgess Hill, Sussex, Campbell approached them: "We had a bit of a romp, had the music on, laughed around a bit, then Donald said to Lew and me, 'Now that you're together, how about designing me a boat?'"[277]

Originally the new craft was to be a Rolls-Royce Griffon-engined proprider, but after Cobb's accident the Norrises quickly sought out Reid Railton for advice about how to develop Campbell's all-new *Bluebird* into a jet-powered hydroplane for the outright record.

At first Railton was hesitant, but once he had decided that he would assist, he immersed himself in a more detailed analysis of the *Crusader* crash in order to ensure that the young newcomers went in the right direction.

The data provided by Railton from the *Crusader* episode gave Norris Bros a massive head start. Through his friendship with Railton, Ken took the art of high-speed design to an entirely new level but always acknowledged his debt to Railton. Ken and Lew expanded Railton's grisly analysis of the forces that had destroyed *Crusader*, determining that their new boat would have to be an all-metal hydroplane. It would become the single most successful boat ever in water speed record history, taking the outright record an unparalleled seven times for Britain between 1955 (202.32mph) and 1964 (276.33mph) before Campbell died on Coniston Water on 4 January 1967 having exceeded 327mph.

Ken Norris: "I got on with Donald straight off, he was a friend more or less the whole way through, because I got to know him very well when we were analysing *Crusader*'s crash. We worked overnight, in the dark, in the office above the garage at his house, and he would sit there smoking his pipe, watching every detail. He was working the projector, frame by frame, Lew was giving me some figures, and I was writing them down. Reid had provided film, photographs and calculations to us, and he was always a guiding hand."[278]

The Norris brothers later created the *Bluebird CN7* car with which

CRUSADER

From
Donald M. Campbell
Telephone Horley 1388
Telegrams Camariner, Horley, Surrey

The Drive,
Horley,
Surrey

21st July, 1952.

Dear Lord Hives,

We are at the moment engaged on the design of a successor to 'Blue-Bird' and our primary object is to challenge America for the Harmsworth Trophy. As you know, they have held this prize consistently since 1920, despite many attempts to wrest it from them.

There is no doubt in my mind but that John Cobb will handsomely succeed in recovering the straight-away World's Record with his new jet craft 'Crusader', a record, incidentally, which Sayres has just raised again to 178 m.p.h.

Cobb's new boat is neither intended nor eligible for any other event, and the view is widely held that it would be an excellent achievement for British prestige if another British craft could, more or less concurrently, succeed in recovering the 'Harmsworth'.

I do feel that this view is fully justified for, of course, the Americans are holding nearly all the major world honours on water.

From the personal standpoint, the old 'Blue-Bird' was the first British boat ever to win the Oltranza Cup from the Italians, and the prospect of her successor recapturing the 'Harmsworth' is an objective worth striving one's utmost to attain.

This endeavour entails a very grave responsibility, for failure would be disastrous, and it is from this point of view that I dare not contemplate the further use of the 'Rs'. These engines have now seen most arduous service and, having done more for British racing prestige than any other unit, deserve most honourable retirement.

-continued-

In this letter to Lord Hives of Rolls-Royce, Donald Campbell seems to be resigned to the fact that with Railton designing Cobb's water speed record contender, he would have to find a different challenge for his new *Bluebird*. Author's collection

FROM THE ASHES

Campbell took the land speed record. Typically, they had insisted that the car had to obey the rules of the time and, try as they might to be different, they ended up with a shape that looked uncannily similar to Railton's pre-war creation for John Cobb. When Campbell crashed the car at more than 300mph on the Bonneville Salt Flats in 1960, it said everything about the integrity of its design that he survived what was then the fastest accident ever to have occurred on land. When the car finally took the record, achieving 403.10mph on Australia's Lake Eyre on 17th July 1964, the only major change to the design was a tall rear-mounted tail fin, something that Railton said his own land speed design, and the *Bluebird*, did not need. Ken Norris and Reid Railton continued to exchange information and ideas right up until Railton's death in 1977.

We can be fairly certain therefore, that Railton would have kept nothing back from the Norris brothers. Ken was staggered at the depth and detail of Railton's analysis, and when he realised how the techniques at his disposal were so much more advanced, he felt it only right that Railton should be "sitting next to us looking at the screen".[279] Alongside the brothers and Railton were Professor Tom Fink (the aerodynamicist on the *Bluebird* project who went on to work with later record holder Ken Warby) and Bill Maloney.

Maloney retained Railton's analysis of *Crusader*'s accident and a copy was also acquired by Ken Wheeler, a highly qualified designer who worked under Ken Norris at Norris Bros and who had a talent for high-speed boat design. Both men assisted the author in more recent research into the accident and this can now be published for the first time.

Although the basic techniques of superimposing datum lines onto period film footage and making innumerable calculations have not changed since *Crusader*'s time, modern technology has allowed film stabilising and enhancement to reveal more. Although the data and information obtained from more recent analysis has confirmed what Norris Bros and Railton had deduced by sheer hard work and number crunching on slide rules, computers have allowed more detail to be added.

In bald terms, the differences between the initial conclusions and more recent analysis are small but telling. Using the BBC footage, it is evident that John Cobb was subjected to intensely violent forces, extreme enough to show that by the time *Crusader* took her final dive, he was probably already dead, or at the very least unconscious. Freeze-frame study quite shockingly reveals that Cobb's head was thrown against the steering wheel with considerable force

Stills from the Norris Bros analysis of the accident: in the second frame, the plume of spray from the cockpit area is clearly visible; the angle of the wakes encountered is fairly oblique, reducing their strength on impact. *Norris Bros archive*

well before the final 'dive in'. He had developed the habit of jamming himself into the cockpit of his vehicles rather than use seat belts and also relying on the foot-operated throttle rather than the hand-operated alternative. Such were the vertical forces that Cobb was almost thrown out when *Crusader* hit the first wave, the resulting 'drop' bringing his head down and forward against the wheel with more than enough force to render him unconscious, which would explain why the throttle remained unaltered thereafter.

The film also clearly shows how, even though the stern-mounted sponsons were riding extremely well, the single forward shoe was rising and falling, as shown by the spray pattern that can be seen travelling forwards and backwards along the side of the 'wedge'. This is indicative of a change to the angle of the forward planing shoe, the area that was subject to constant strengthening during the trial runs, and that would have compromised fore/aft stability.

Also evident is something, very probably the rudder, folding out sideways from the spray. Curiously, the accident that befell *Crusader* has been replicated consistently both in computer simulations, and in scale, with powered models, by the simple expedient of hitting hard helm while at speed. Indeed, master modeller Ernie Lazenby successfully replicated the crash both with and without deliberately creating rough water. Sadly, the audio of all the film available cannot be accurately timed to the visual aspect, but this rudder-like shape seems to become noticeable just after the loud crack that was heard.

Bill Maloney, on seeing some of the computer model 'crashes' and the enhanced footage, was moved to comment: "If you look at the position of the rudder and the construction, size and weight of the later version, and its attachment to the rear of the front shoe, it doesn't take a great deal of imagination to think that due to the oscillations the rudder could have been ripped off the boat due to the continuous resistance changes due to its varying depth in the water and the hammering the shoe was taking at the same time."[280]

Added to this are Cobb's comments after the rudder had been changed from the flat sheet version to the larger cast item: at speed, he judged that *Crusader*'s handling had deteriorated and felt as if the wheel (tiller) was constantly rocking in his hand, most likely because of a phenomenon known as 'rudder flutter'. This trait can be a consequence of various factors in a rudder's design, namely excessive weight, lack of symmetry, exaggerated size, or a section shape that allows alternate attachment and detachment of flow on one side — or a combination of all of these. Flutter in any form happens in cycles, rising to a high frequency and setting up tremendous vibration. Once a rudder flutters, higher loads are imparted to the surrounding structure. In *Crusader*, this vibration would have been horrible for the structure and the driver, and would have been transmitted back to the steering wheel, much as Cobb described.

Those who studied the BBC footage were all struck by the extreme vibrations displayed, as supported by various statements and observations of the time. As Maloney added, "Rudder flutter, yes, this would have added extra fatigue to the structure."[281]

We must also refer back to Railton's thoughts when the position of the rudder was decided. While he agreed that its position was the only available solution, he warned that it would be difficult to predict how much control would be available, with "too little deflection creating an uncontrollable boat, **and too much causing the craft to capsize** [author's emphasis]."[282]

It cannot be confirmed one way or the other that the rudder failed or parted company with the hull, but it does seem likely because it was replicated in all analysis test runs, both virtual and physical, simply by selecting full helm either way. What is certain, however, is that questions surround the structural

FROM THE ASHES

The two types of rudder with which Crusader *ran at Loch Ness: the Thomson & Taylor original (above), which was a fabricated, adjustable flat plate, and Vosper's larger cast version that was subsequently fitted. There was also another Vosper rudder, larger again, that may have been fitted to* Crusader *at the time of the crash.*
© *Castrol Ltd and courtesy of BAE Systems (formerly Vosper Thornycroft)*

integrity of the forward planing area, to which the rudder was attached.

Analysis of the footage with the applied datum lines shows that Cobb experienced a longitudinal pitching oscillation of such intensity that the overall craft was unable to constrain or absorb it, and the most vulnerable part of the design — the single front planing surface — was unable to withstand the punch loads being applied to it. The analysis shows that the craft suffered violent pitching oscillations of 5–6Hz, initiated by a swell outside the measured mile, amplified upon encountering further rough water, and leading to a further and larger distortion of the front 'wedge', either by an alteration to its attitude or in conjunction with the failure of the rudder mount. Such instability was shown to be related to the height of the centre of gravity above the planing surfaces and their inherently large moments therein, plus factors relating to the 'spring rate' of the planing surfaces and the high thrust line.

CRUSADER

With *Crusader*'s bow high out of the water, at the rear of the forward shoe a dark shape can be seen where the rudder is mounted, creating a large spray pattern.
BBC Film Archive

Tom Fink once said the following about high-speed accidents on water: "There was always this, 'they hit a submerged railway sleeper just below the surface', well, where did that come from? It supposedly happened to Donald [Campbell] in 1951, but to me, that just seemed too good to be true, too easy. So they [Norris Bros] looked at the news film of the Cobb accident. It was displayed on the wall with the lights out so that we could see really large pictures of the boat. It was quite obvious, frame after frame, that it was pitching heavily. And then there was just a little spout of water visible on the upper deck, just in front of the windscreen, which showed the first break must have happened down at water level, letting water come up from underneath. And that said to us that it must have been a structural failure associated with the oscillation period. So the main thing was that we were able to analyse it more closely and work out what the stresses must have been which caused the breakage. Of course, it was mainly only a wooden structure."

Most of the period analysis assumed that *Crusader* was 31ft long, but the Vosper 'General Arrangement' drawing indicates that the length was 29.25ft. Assuming an all-up weight of 6,500lb (there is no log of how much fuel she was carrying and we cannot know how much water she had shipped in the transition stage), the crucial frames of the BBC footage (at 24 frames per second) run from frame 179 (7.46 seconds before final dive-in) to frame zero (the moment of final dive-in). Ken Norris, Bill Maloney and Ken Wheeler all agreed that there was a centre of gravity loading of 40,800lb at frame 179 and suddenly the much higher figure of 93,000lb at frame 176 (0.13 seconds later). As the centre of gravity was behind the forward planing shoe, the loads on the structure above the shoe would have been considerably more than at the centre of gravity, so this moment would have been the critical time. In his 1955 analysis, Ken Norris suggested a load of above 28G, about the level that Railton had said the boat needed to be designed for, whereas Peter Du Cane had suggested that 11G would be sufficient.

Lorne Campbell has trouble reconciling this discrepancy: "PDC [Peter Du Cane] should have believed RAR [Reid Antony Railton] here. I don't understand why he didn't. It looks as if he failed to take all of the factors into account. I suppose RAR was doing his own calculations while PDC probably delegated them to one of his team. Maybe PDC didn't check them?"[283]

Having said that, it is important to understand that all of those involved in the design of *Crusader* were working very much in the dark, in a field new to them, and that *Crusader* fell foul of the little-known effects of a dynamic instability, elegantly addressed by Ken and Lew Norris, and recently confirmed by Bill Maloney, Ken Wheeler and the author.

It is also possible to argue that even if *Crusader*'s front planing shoe had been strengthened sufficiently, such extreme forces on a single, flat surface would have led to failure. And even if Cobb had been able to run on perfect water, eventually a speed would have been reached at which the oscillation was inevitable, and no record breaker would hold back if more speed was available.

As for his speed, Cobb was accelerating hard at that point, but this is where the flat speed/resistance curve worked against him. Bert Denly, although certain that *Crusader*'s fateful run was merely a high-speed trial, was equally convinced that Cobb was putting on a show and, at the highest speed he had yet experienced on water, realised he was in trouble. Having lifted from the throttle, he found himself continuing to gain speed through the violent pitching until it rendered him unconscious, just before a key part of the boat catastrophically failed. Had there not been a structural failure at that moment, *Crusader*'s limited dynamic stability would have eventually manifested itself anyway and the craft would have become uncontrollable in pitch, at a higher speed and with the same disastrous consequences.

Without implementation of the full remedial work that had shown itself to be necessary for the structure around the front planing shoe, the combination

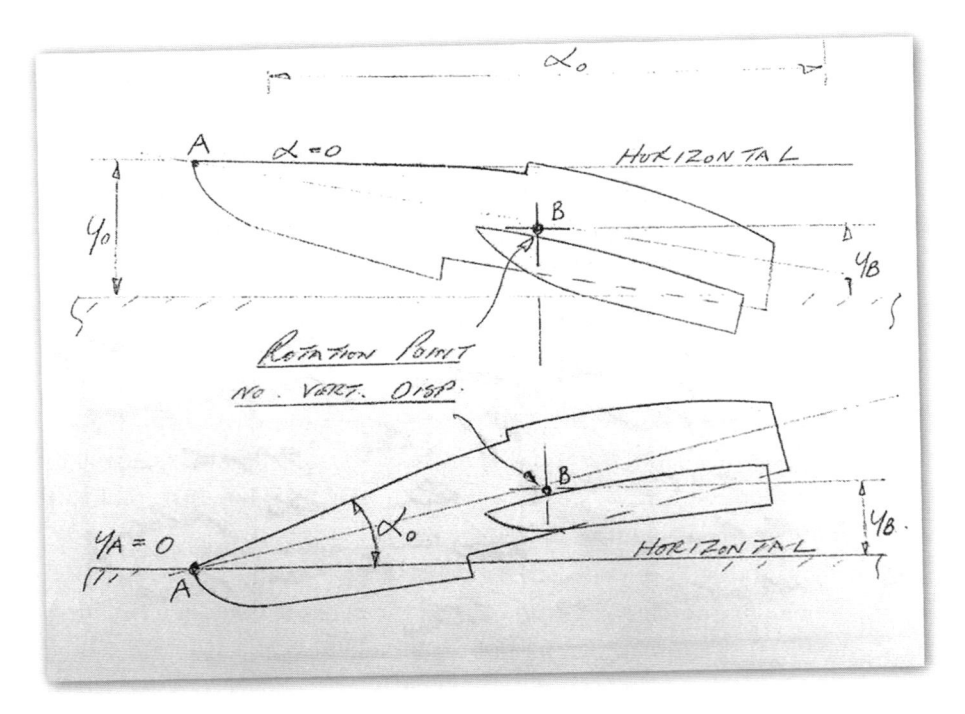

Sketches from the Norris Bros analysis into the *Crusader* accident show fore/aft pitch data (above) and a graph of oscillation patterns. *Norris Bros archive*

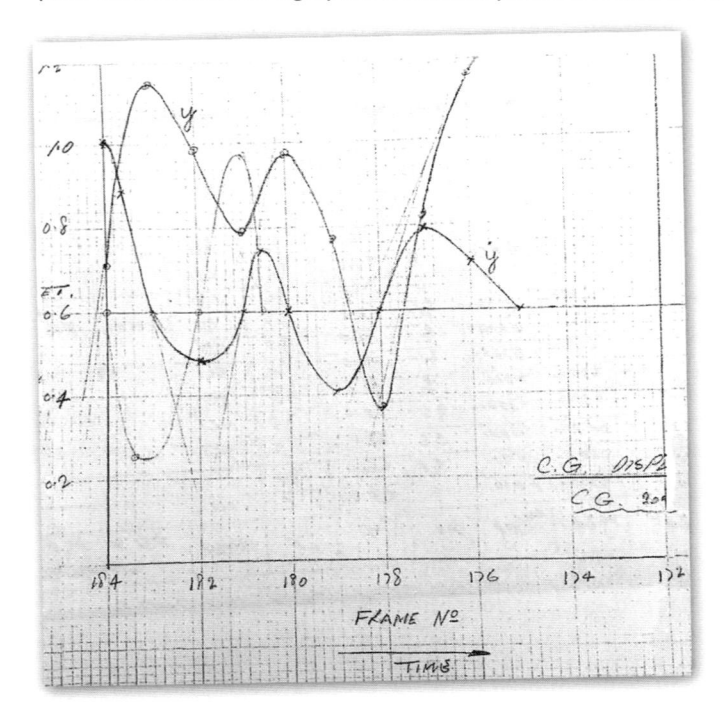

of weakness in this area and the possibility of rudder flutter adding strain to the shoe's mount was probably only ever going to have one outcome. In this regard, it should be added that if the forward shoe had been vee'd, it might have survived the final dive in, albeit damaged, and had the stern shoes been similarly vee'd, perhaps the bow would not have been 'tipped in'.

Lorne Campbell, however, doubts that any vee could have saved the craft:

> If the shoe was collapsing, the boat would still have dived in, whatever the shape. The vee might have reduced loads enough for a similar structure to survive without initially collapsing, however.
>
> The lack of vee is odd. Du Cane appears to have had a mental block in appreciating the loading on the forward step. Maybe his concern about porpoising took over.
>
> The kick-up from the rear sponsons coupled with the bow dropping naturally due to the thrust/drag couple does mean that the boat was somewhat bow-down when landing, but the forward surface had a high angle of attack so I don't believe it actually landed at a negative angle. If, however, the impact with the second wave had finally made the forward shoe cave in, then the transom of the forward step would have acted like a very large water brake which would, for an instant, put enormous drag on the shoe area with consequent enormous deceleration on the craft at the shoe level. With both the momentum from the high CG [centre of gravity] and the thrust line being higher, the craft would have started to pitchpole forwards. But then the pressure of the water entering through the step area would immediately start to blow that area of the boat apart — basically slicing the hull apart from the bottom upwards. This shows why the bow portion is seen to bend upwards in the film — it was being snapped off upwards with the split starting at the step.[284]

Ken Wheeler did not doubt the weakness of the front shoe support: "Ken Norris analysed the *Crusader* footage on the last run which recorded a pitch oscillation of +/– 1.5 degrees at about 5Hz. From the factory drawings, in my view the load path provided by the keel type structure used did not match up to the forces involved. Why Du Cane did not put another hoop around the hull at the front shoe as he did for the sponsons defeats me."

Of course, Reid Railton had envisaged an alloy hoop in the hull at the front planing shoe to serve various purposes: to give a 'punch load' path into the main hull, to provide a substantial mount for the rudder assembly, to offer a load-bearing hoisting point, and to add to the pilot's safety. It is clear from archived letters that Peter Du Cane or Vosper assumed at the outset that *Crusader* would be launched from a slipway and therefore decided that the forward alloy frame for the rear of the forward step and rudder was not required, disregarding the fact that Railton had done the stress calculations for that area including the alloy hoop. It is also clear that Du Cane, upon learning that the craft was to be crane-launched, realised that the alloy frame was most definitely required, such that when told of the need for a lifting eye forward of the cockpit he informed John Cobb that the area of the hull where the hook was to be mounted should be "considerably strengthened". But by this time, as confirmed by the dates of relevant correspondence, Vosper had already framed up the main hull and therefore would have needed to add a plywood frame forward of the cockpit in order to accept the lifting eye.

In Vosper photographs of the hull being built, and more specifically of the vertical plywood frame/bulkheads, it is apparent that the frame that lines up with the forward lifting eye and the rear of the forward planing shoe is *not* cut from a single sheet of plywood, as would have been desirable for strength and integrity, but contains a section let in at that critical point. And what of the cockpit opening and the jet air intakes? Looking back to the 'General Arrangement' drawing, it is evident that the forward edges of the air intakes line up with the rear of the shoe, yet to accommodate them the single plywood frame would have had to be cut, as would the cockpit opening, introducing at least two further weak spots in the same crucial area. Added to that, this same frame is the one to which the lifting eye was bolted. Compounding things even further, the main keel girder in this very area had a bolted lap joint.

Additionally, the front shoe, which Railton had originally specified as forged alloy but had been drawn up by Vosper as an aluminium/wood composite, had actually become in reality a single alloy sheet of the same original dimensions. Du Cane seems to have believed that most of the loads would be on the rear shoes, even though Railton had pointed out to him how much higher the forces would be on the single forward shoe — and this proved to be the case. But, as has been stated previously, all involved were working in new territory, with limited access to even some basic materials due to shortages, with only the rudimentary technology available at the time, and with convoluted communications.

FROM THE ASHES

Two close-ups of the hull during 'framing up'. These illustrate where the continuous bulkheads were cut not only in the cockpit and air intake areas but also at the rear of the planing step, which is clearly lower and made from a different piece of wood.
Courtesy of BAE Systems (formerly Vosper Thornycroft)

CRUSADER

The reverse three-point design produced huge loads on the single front plane, loads not expected or considered by some during the research and design. It is very likely that the front planing shoe did partially collapse, punching into the forward shoe structure. When the rudder began to flutter, creating severe oscillation, it either tore off or caused the hull to tear upwards from the shoe. Either way, the vertical transom at the stern of the step assembly would have become exposed, presenting a huge surface area that would have acted as both water brake and scoop, creating the vertical spurt of water seen in film footage just as the hull tripped in. The bolted lap joint would have failed, severing the bow assembly along the one, incomplete, weak plywood bulkhead, like bending a banana in the middle. Indeed, as can be seen through the spray, the attitudes of the bow and the stern do not match, and, with the stern high out of the water, the bow — complete with pitot tube — can be seen pointing skyward.

During his research for the construction of *Bluebird K7*, Tom Fink, having watched all the available film of *Crusader*, pinpointed a slight burst of spray just in front of the windscreen, *before* the nose began to dive in. This suggests a structural failure, technically below the high-speed-running waterline, indicating that perhaps the alloy sheet planing surface had indeed punched up, taking the vertical supports up into the hull with it.

It can only be assumed that the shoe distortion at high speed at this very area of the hull cannot have been helped by the plywood frame also being used as a lifting point, nor that the 'punch loads' were to a flat surface as opposed to one with some degree of vee to them.

Indeed, in March 1970 Peter Du Cane stated in a letter: "One could say the *Crusader*'s forward step should have stood up to going into the skip of the large waves (for this type of boat) but I knew the step was a bit weak, and it had shown signs of it when we ran previously, and I asked John Cobb to let me have her for a month, at our expense, to strengthen this rather doubtful point."[285]

As Vosper had offered to pay for this remedial work, the company must have felt a degree of responsibility, especially when one considers that doubts had been cast early in *Crusader*'s trials when Harry Cole referred to some slight distortion after her first trial in the sea, something confirmed in another Du Cane letter dated 6th June 1970: "With these record breakers, models will take one quite a long way, but not by any means the whole way. I envisaged 300mph odd, but I would have changed the planing surfaces, 'V' them, to avoid the rough ride."[286]

FROM THE ASHES

These two stills from the Movietone News film are separated by no more than a quarter of a second. The first still shows a small spurt of water beginning to erupt from in front of the cockpit, resulting from water entering the hull below the water line. The second still shows the plume engulfing the cockpit as the bow goes in.
Movietone News Archive

Through no single individual's fault, *Crusader* was inherently uneasy in pitch due to the high centre of gravity and high thrust line inflicted by the weight and large diameter of the Ghost jet engine. Imagine being at the top of a double-decker bus with a large, heavy, in-line internal combustion engine also mounted at that level: when the engine is revved, there would be resultant movement from side to side due to torque reaction. On *Crusader*, this effect would have been fore/aft, as opposed to in yaw (side to side).

Furthermore, the position of the rudder was statically unstable because it was in front of the centre of gravity, so if a course deviation developed at high speed, or the rudder presented an excessive profile to the direction of travel, the resultant instability in yaw would be as capable of overturning the craft as instability in pitch. Imagine making a sudden change of direction at high speed in a three-wheel car such as a Reliant Robin (with the single wheel at the front): because of the height of the car's centre of gravity, a large steering input at speed would threaten to turn the vehicle over.

There is also the possibility of an additional contributing factor to *Crusader*'s final plunge — the design of the bow section itself. Peter Du Cane had exhibited a degree of tunnel vision in his determination to eliminate porpoising while also being aware that the vessel had to lift onto her plane as quickly as possible. To

In this still from Jock Gemmell's Pathé News film, showing *Crusader* as she dives in, the severed bow complete with pitot can be seen, apparently separated from the main hull. *Pathé News*

that end, the angle of the forward planing shoe was relatively steep, but it also had to merge in with the bow, giving it a very definite, steep, upward curve.

We must now refer back to the Marine Aircraft Experimental Establishment's response to the design when Du Cane requested the loan of an accelerometer. The MAEE stated that the planing surfaces had to be vee'd at all costs. The MAEE's commandant, W.C. Winn, had once been part of the High Speed Flight, the wing of the RAF tasked with winning the Schneider Trophy, the famous competition initially run for flying boats and then, as speeds rose, for seaplanes. The designers of these aircraft encountered all of the problems that faced *Crusader*, and more, because they were seeking speed on water as well as in the air. As take-off and landing speeds were relatively high, as much attention had to go into float design and hydrodynamics as into aircraft design and aerodynamics.

The MAEE's concern with *Crusader* was not just the need to soften her ride across water by veeing the surfaces, but also to reduce the Coandă effect along the 'buttock lines'. The Oxford English Dictionary defines a buttock line as, "(*aviation*, *ship-building*) A curve indicating the shape of an ***airfoil*** [*author's emphasis*] or nautical equivalent in a vertical plane parallel to the longitudinal axis of the craft or vessel." The Coandă effect, named after Romanian inventor Henri Coandă, is the tendency of fast-moving liquid or gas to remain attached to a curved surface.

FROM THE ASHES

As previously stated, Railton had deliberately specified that the rear sponsons should have gentle curvature to aid getting 'over the hump', while also keeping the planing surface as shallow as possible. On the bow, however, the buttock lines formed a very severe curve, which, although ideal for getting over the hump quickly, could cause problems if the angle of attack were altered.

This problem had affected the Supermarine S6 and S6B seaplanes used in the Schneider Trophy. These aircraft were known to be very tricky on take-off and landing, tipping in nose-first if the aircraft's attitude was allowed to change. Indeed, Lieutenant G.L. 'Monty' Brinton was killed in 1931 when his S6B tipped in, in flat-calm conditions, which led to considerable but unsuccessful head scratching. It was some years later that the MAEE realised that the cause was the Coandă effect acting on the curved under-surface at the front of each of the aircraft's floats.

The Coandă effect has also been used very recently in Formula 1, by harnessing engine exhaust to accelerate airflow around the curved surface of a rear diffuser to create downforce. In water, the effect is no different. If one places a spoon against the flow of water from a tap, the water can be seen curving around the bowl of the spoon, clinging to the surface, much like air over an aircraft wing. On an aircraft, this causes low pressure, and the wing is 'sucked' into this low-pressure area, to fill the void. Turned upside down, on a racing car for example, this pulls the car downwards, improving grip. With *Crusader*, the Coandă effect was neutral when she travelled at speed at a consistently level attitude, but on 29th September 1952, as we have seen, she was not running anywhere near level.

Crusader was observed running anything but smoothly before encountering the wave pattern. Whatever the source of this wave pattern, it had a major influence on the outcome. The bow wave of a small, relatively fast boat leaves a persistent, diverging wave of about 20 degrees to the craft's centre line, and in flat-calm conditions, can last a considerable time without any damping element, other than that of the water's own viscosity. With Loch Ness's steep sides, waves might be reflected back, at an angle of approximately 60 degrees toward the centre, then back and forth several times, as confirmed by long-time loch boatman Hugh Patience. Assuming the craft that created the wave pattern had a speed of precisely 12 knots (more than the *Maureen*), the reflected wave pattern would continue to travel at an angle of 60 degrees at 0.433 knots, and would continue to advance along the loch at about one fifth the speed of the vessel that created the waves.

CRUSADER

Two rear views of the dive-in, both from the BBC footage. The long-distance shot is just before water first burst through the top of the hull. The next image, taken from the film for use by newspapers, is a quarter of a second later, and shows the force of the water from within the hull. *BBC Film Archive/ Associated Press*

Analysis of the Movietone News film footage, shot at 16 frames per second and covering the final 16 seconds of the run, shows no definite structural collapse until the bow section dives in. Immediately before the final plunge, it shows the bow high enough for the entire front planing shoe to be clear of the water, then less than 0.25 seconds later the bow is so low that water is spraying from almost halfway up the steep upward curve of the bow/planing surface. It is then only 0.0625 of a second until water is seen gushing upwards from the cockpit opening after the hull has been compromised.

When the hull's angle of attack altered, especially when the rear shoes hit the waves that the bow had skipped, the steep, aggressive buttock-line curves magnified the Coandă effect, pulling the bow down and into the water, which became increasingly dense as the hull dived in. In whatever order the hull punctured or the rudder detached (fully or partially), it all led to the shattering of the weakened hull.

As an indication of just how badly *Crusader* was vibrating, analysis of David Low's BBC film shows that from frame number 184 (counting down towards the final dive in), a datum point just behind the cockpit (in line with the forward sponson arm and alloy ring structure) rose by eight inches, fell by four inches, rose by three inches, fell by eight inches and rose by eleven inches — in under a second.

Philip Mayne, one of the timekeepers, had the advantage of being in the middle of the course at water level, but looking down at his watches as he operated them. At the time he recalled what he heard: "The boat was going magnificently, and must have touched at least 215mph. I could hear the rush of water, and an irregular thumping, like a conventional boat would make over waves. Cobb completed the measured mile at 206.8mph, but about 150 yards from the post there was an incredibly loud pistol-like crack, his engine shut off momentarily. He opened up again just before he reached the end of the mile and was going full bore. When I looked up from the stopwatch the boat had vanished, leaving only floating wreckage and silence."[287] Again, there is no reference to an explosion.

When one considers all of this, the fate that befell *Crusader* was a typical accident, the culmination of many factors, some avoidable, others not, all coming together at just the wrong time, and ending in a way that none of those involved deserved.

Lorne Campbell: "I think it was all most unfortunate. There is no reason why *Crusader* shouldn't have worked — basically it did — but the structure wasn't up to it, and that was mainly because Vosper didn't seem to believe or take notice of the design loads that Railton said it would have to withstand."[288]

Then, just as this book reached completion, a previously unpublished eyewitness account surfaced. It was written by R.A. Bruce, captain of the official observers' vessel, the *Maureen Mhor*. Bruce was a trained observer who had served in that capacity during the war and had good knowledge of boats and the loch. As he was very close to the accident and had an uninterrupted view of it from the *Maureen*'s upper bridge, he had the best vantage point of anyone in the vicinity that day.

It was the *Maureen* that was seen as the cause of the three waves that travelled into the path of *Crusader* approximately halfway through the measured mile. Bruce observed the three waves but had his own views about their source and wrote his thoughts down within days of the accident:

> My recollection of the accident is that *Crusader* struck a series of three waves which I would estimate to be about six inches in height. It is

difficult to say how these waves occurred except that they must have rebounded off the shore and were possibly caused by the wash of the Jollyboat, although I would say that at least 15 to 20 minutes elapsed after the Jollyboat passed this position and the accident occurred. These waves were, in my opinion, 'running up' the loch from South to North [this is probably a generalisation as it would be more accurately expressed as south-west to north-east].

I distinctly remember when *Crusader* passed my position that although the surface of the water was like glass, the *Crusader* was **vibrating so much that Mr Cobb appeared to be bouncing up and down in his seat** [*author's emphasis*]. (My engineer remarked on this point later that day).

After striking the first wave the engine was momentarily out and the boat 'bucked' violently once or twice, then the engine was heard to accelerate, the stern appeared to lift.[289]

Bruce made no reference to comments that the *Maureen* herself had been the source of the three waves, but did note that they must have been created by a craft moving faster than the *Maureen* could go. And, rather than repeating some comments that *Crusader* had been running "perfectly", he clearly saw the violent oscillation that later analysis determined.

More telling is Bruce's estimate of the wave height as no more than six inches. When looking back at discussion between Railton and Du Cane, Railton had estimated that a wave of six inches would place a load on the forward shoe of some 11G, and suggested building that part to withstand twice that force. Du Cane's response had been this: "Can you honestly believe that this boat would not be able to pass over a 3" wave? We seem to be overdoing the theory a bit..."[290] When one also considers that Railton had stated that *Crusader* should be run in "flat calm only", we can see the importance of this matter.

One other observation from Bruce is noteworthy. In a conversation with Bert Denly, he said, "We got a radio message to stand down, and everyone was to hold station. So we had waited about five minutes, when we looked up the slope to the radio car, and saw someone waving and pointing to the west, and the car drove off. As [Arthur] Bray couldn't get confirmation from base, the pressmen decided it had been called off, it had to have been if the radio car had left its station, so they badgered us to head off."[291]

FROM THE ASHES

In this hand-written letter card dated 25th March 1970 and sent to author Kevin Desmond, Peter Du Cane conceded that he knew the front shoe was "a bit weak" before *Crusader* had run, and that he volunteered to strengthen this "rather doubtful part" at Vosper's expense. *Courtesy of Kevin Desmond*

CRUSADER

John Cobb's Jet-Propelled 31-ft. Boat "Crusader"

Length . . . . 31 ft.
Beam . . . . 13 ft.
Static Thrust of De Havilland
    Ghost engine . . . . 5,000 lb.

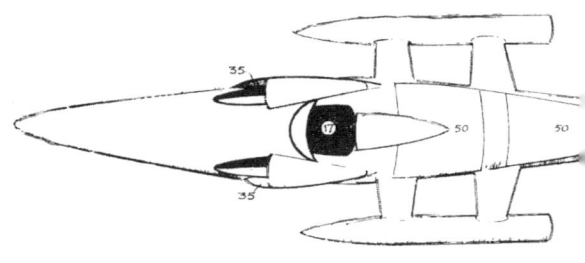

This cutaway drawing was done by *Motor Boat and Yachting* with the help and approval of Vosper. It clearly shows the two alloy ring bulkheads, but also claims that the bottom of the hull is all metal. The drawing also illustrates how the rudder post was fixed to the vertical wooden bulkhead rather than being part of a one-piece bulkhead. It does not show the lifting eye forward of the cockpit nor how the intake trunking passed through the bulkheads.

Motor Boat and Yachting

FROM THE ASHES

1. Main girder (port).	16. Stressed skin (double diagonal plywood).	33. Fuel tank.
2. Main girder (starboard).	17. Pilot's seat.	34. Air trunk—outline.
3. Main girder alters section.	18. Throttle (foot).	35. Air guide and spray guard.
4. Bottom stringers.	19. Throttle (hand).	36. Port float.
5. Gunwale or main stringer.	20. Instrument panel.	37. Starboard float.
6. Gunwale or main stringer alters section.	21. Air speed indicator.	38. Metal bottom.
7. Thrust beam.	22. Tachometer.	39. Metal stringers.
8. Birch ply frame (No. 23).	23. Air drogue release.	40. Fin.
9. Moulded nose.	24. Drogue stowage.	41. Cheek plate.
10. Step.	25. Drogue tackle eye.	42. De Havilland Ghost engine.
11. Transom.	26. Square chine.	43. Jet pipe.
12. Steering drop arm, drag link, tiller arm and rudder stock.	27. Moulded bilge member.	44. Compressor turbine.
	28. Forward cantilever arm.	45. Combustion-chambers.
13. Rudder.	29. After cantilever arm.	46. Compressor.
14. Starboard bracket at step (port bracket similar).	30. For. strong beam (portable).	47. Starter motor.
	31. Aft. strong beam (portable).	48. Vacuum pump.
15. Cockpit carline.	32. Strong beam securing plate.	49. Oil filter.
		50. Engine cowling (portable).

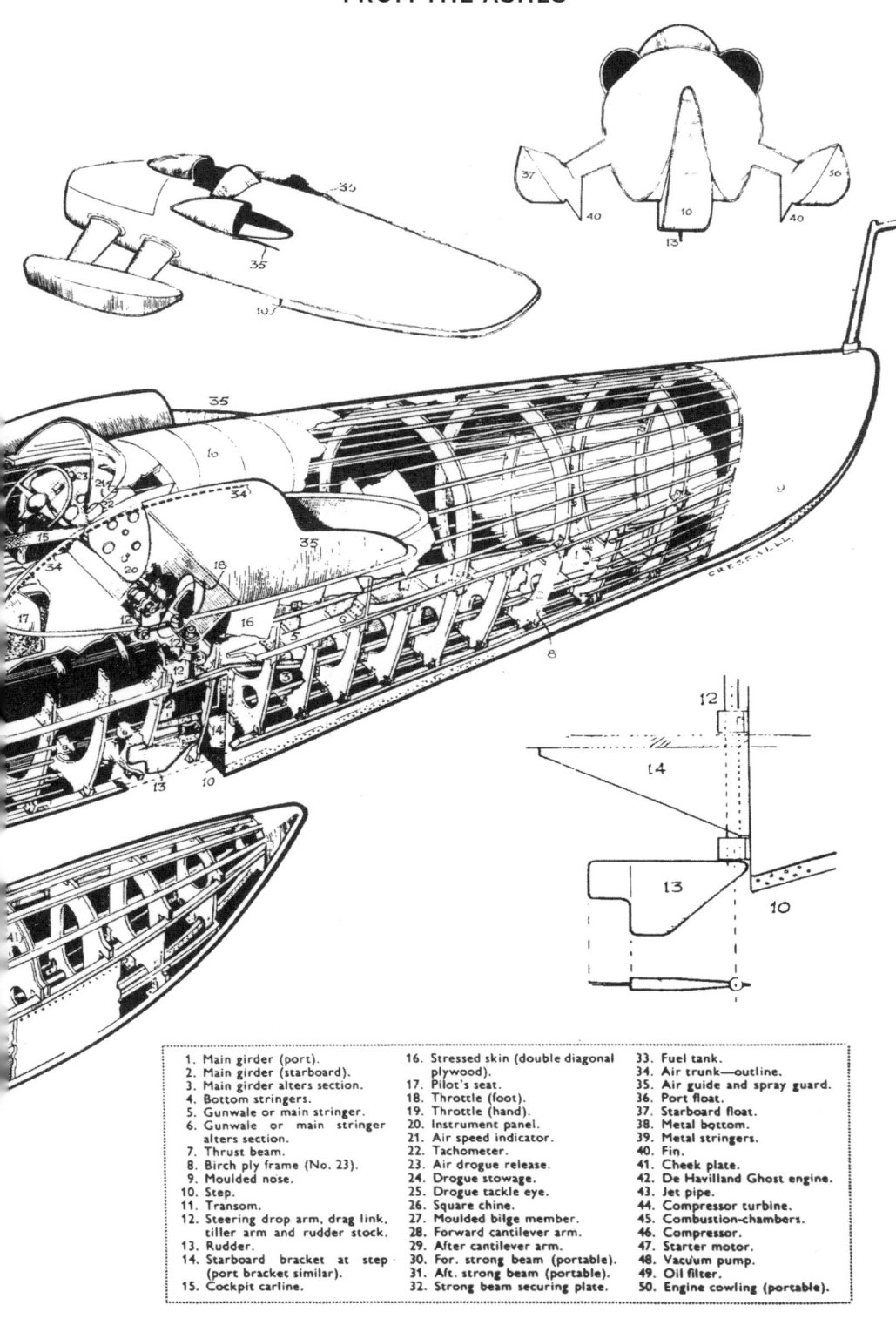

CRUSADER

In summary, *Crusader*'s downfall — as in all accidents — came about through a combination of elements, which through correct crash analysis can point to the root cause or causes.

From a study of the films, photographs and witness statements, model tests, both real and virtual, we can pinpoint nine contributory factors:

1] Disturbance of the water surface.
2] The lack of dihedral (vee) on all planing surfaces.
3] Inadequate frame strength at the rear of the forward planing shoe.
4] A questionable mix of construction and repair materials.
5] Rudder size/symmetry and failure.
6] Coandă effect at the bow assembly.
7] A flat speed/resistance curve.
8] Height of thrust line and centre of gravity.
9] Fore/aft instability.

Some of these elements, of course, were interlinked, and some were out of the control of all those involved, and leading to these conclusions:

- There was disturbance to the surface of the water that most likely came from the wash of a support boat. Because of the speed necessary to create the disturbance, it was probably caused by the 'jolly boat' or the *Isabelle*, and possibly involved a form of Soliton effect.
- Had *Crusader*'s planing surfaces been vee'd, this disturbance might have had little or no effect, but no vee was applied, and the large, flat surfaces, especially at the bow, took high impact forces.
- Such forces might not have had an effect, but the hull framing at the very point where these vertical forces occurred was compromised. There was no alloy ring to transmit the forces into the main body of the hull. The structure was not one unbroken bulkhead but a frame, with joints either side of the planing wedge mount, and gaps created by the engine air intakes and cockpit opening, and with added stress from having the weight of the vessel on that very frame via the lifting eye, directly above a bolted lap joint in the main longitudinal keel members.
- Added to this, the insert between the frames at the rear of the planing wedge also had the rudder stem attached to it. This rudder was larger

FROM THE ASHES

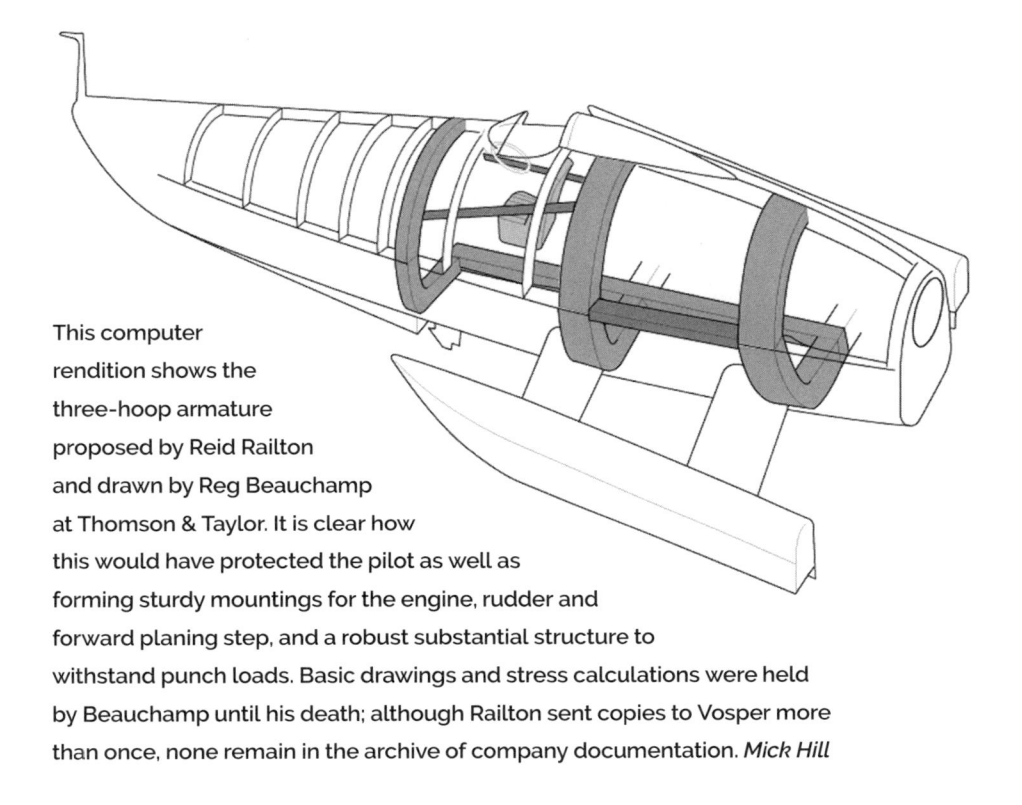

This computer rendition shows the three-hoop armature proposed by Reid Railton and drawn by Reg Beauchamp at Thomson & Taylor. It is clear how this would have protected the pilot as well as forming sturdy mountings for the engine, rudder and forward planing step, and a robust substantial structure to withstand punch loads. Basic drawings and stress calculations were held by Beauchamp until his death; although Railton sent copies to Vosper more than once, none remain in the archive of company documentation. Mick Hill

and heavier than the one originally intended and originally fitted — and possibly less than symmetrical.

- Something in this area was heard to fail: either the rudder support or the frame that held it, caused by the rudder fluttering, or more likely, the front planing area, milliseconds before.
- To compound this further, the mix of materials chosen for this very point, first during construction and the more so for 'in the field' repairs, was ill-conceived.
- Structural failure in this area would also have altered the angle at which the aggressively curved bow section of the wedge interacted with the water's surface, to the point where the craft's attitude was excessively altered in relationship to the surface. In turn, that could have allowed the Coandă effect to develop in this area, although it is evident, as mentioned, that the hull snapped up and off at the weakened area moments after the shoe gave way, allowing a jet of water to burst in.
- Long before reaching the disturbed water, Cobb would have felt significant change in the boat's behaviour, as indicated by the spray pattern at the

CRUSADER

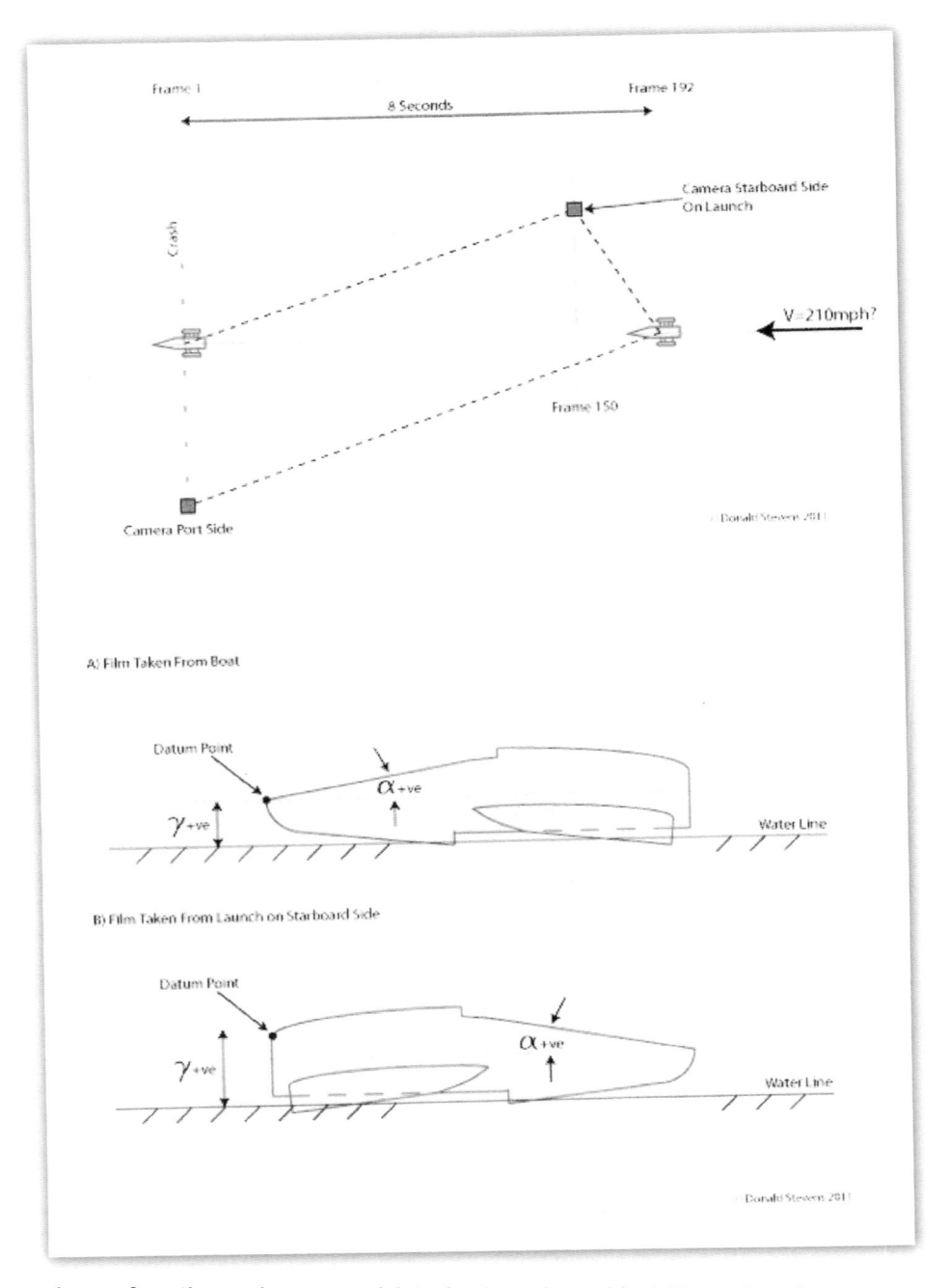

A page from the modern research into the *Crusader* accident. The author, along with ex-members of Norris Bros, computer and physical modellers and noted marine architects, brought modern techniques into a new, in-depth study.
Courtesy of Donald Stevens

bow — but *Crusader*'s flat speed/resistance curve meant that he would have been beyond gaining any useful deceleration by shutting off power.
- Close scrutiny of film footage confirms that Cobb became unconscious during the final seconds of the run. The reapplication of the throttle noted by various observers could have occurred after his loss of consciousness, from the weight of his foot or hand on either of the controls.
- The altered angle of the front planing shoe also magnified the inherent problems forced on the design by the size and dimensions of the engine, putting both the thrust line and centre of gravity too high in the hull, creating fore/aft instability and — just at the wrong time — contributing to the change in the hull's attitude to the water.
- Running when there was water disturbance, however created, can be put down to 'operational error'. Not enough time was allowed for various washes to die down, and in the absence of Reid Railton, who insisted on 'flat calm', *Crusader* was allowed to run in conditions suited to a trial run — which this was supposed to be — rather than a record attempt.
- Cobb's excessive speed for the conditions may have been due in part to him 'putting on a show' for the large crowds, and was made possible by starting further from the measured mile than previously, allowing a longer run up.

The high thrust line and centre of gravity were unavoidable as the engine was really the only one available, but all the other deficiencies in the design and construction could have been avoided. The conclusion must be that no single person should be held responsible for the failure of a futuristic craft that was both prototype and finished article in one, but one has to wonder how things might have played out had the craft been built as had been requested. It seems incredible that, when faced with several expert opinions, much that should have been done was not.

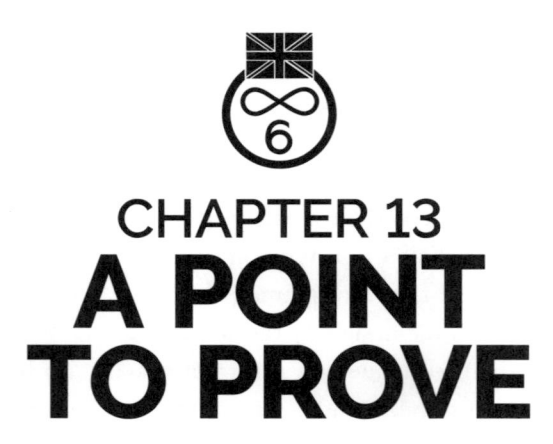

CHAPTER 13
A POINT TO PROVE

Within a few months of the accident Reid Railton commented in a letter: "We'll never know exactly what happened, since the evidence lies in 1,000ft of water, where it is likely to remain."[292]

Railton had left for his California home only the day before *Crusader*'s fateful run, with the intention of returning for a subsequent attempt within two or three months, or the following year. Shocked as he was by the loss of his friend, he allowed himself nevertheless to consider what he believed should have been done differently. This soon led to the idea of building another boat that would redefine his original concept and the thinking that he had put forth, and that he felt had been compromised by the flaws in *Crusader*'s construction.

On 14th December 1952, just over two months after the accident, Railton wrote to Richard 'Dick' Wilkins, Cobb's friend and chief executor of his estate, laying out what he thought might be done next:

> How goes the big executor! I should love to hear all about it but I expect I shall have to wait until I see you.
>
> I have been doing a lot of thinking about the *Crusader* episode, and think that I can now see the thing in true perspective, though nothing can excuse the shocking bad judgement that led up to it.
>
> I have also been thinking a lot about possible future activities in that line, and have come to some fairly definite conclusions as to how to proceed if opportunity should offer.
>
> I must confess that I should quite like to prove to the world that the

A POINT TO PROVE

idea was all right provided it had been executed a bit better.

Are you still interested? I shouldn't be in the least surprised to learn that you have come to your senses but in case you have not I will outline a plan of campaign that I have in mind.

In the first place I think the whole thing should hinge on getting Wakefields' co-operation, as by Wakefields [the company behind Castrol oil] I really mean George Eyston. If we could get him to give us the same wholehearted support that he gave John and have him in from the very start (instead of only at the end) then I think I can see a way of handling the job properly. You probably don't realise what a tower of strength he was to us last summer, and how admirably he compensates for my own shortcomings. You would be far better off with Wakefields and George than with say Shell and twice the money.

The programme might then go something like this

1. I should prepare drawings of the general design over here (at my own expense).

2. Simultaneously I should approach Saunders-Roe through certain private channels, to see if they would be interested.

3. I should come over say in March and together with George try and get an engine out of Rolls-Royce or failing them out of De Havilland's.

4. At the same time I should make definite arrangements with Saunders-Roe (or whoever) and leave them to stew over the details of construction and to produce proposals.

5. I should come over again probably in August and live with Saunders-Roe until the design was finalised.

If no time is lost this programme should result in the boat being ready around July 1954. You will note that it leaves no interval of time for making and testing models. The reason for this is that I think we have enough experience already to make the risk of dispensing with model tests a reasonable one <u>except for the personal risk to the pilot</u>. I should propose therefore to equip the boat with remote control, and to conduct the initial trials <u>of the boat itself</u> without any pilot. Loch Ness would be an admirable place to do this, and I think the equipment could be borrowed almost ready made from the Air Force. When everything was proved O.K., the victim would take a stiff whisky and soda and seat himself at the controls.

> As I had promised I wrote to Howard Robison a few weeks ago merely saying that I didn't think he ought to fool with the idea at all unless he could lose £20,000 without missing it.
>
> I know that you have the reputation of being a very costive [meaning 'reluctant'] correspondent but you might scribble me a line to say what your latest feelings are on the subject. I expect to be at the above address [The Whittier Hotel in Detroit] until the end of January.
>
> I should be interested to hear what sort of final settlement you made with Vospers. I wonder what happened to the 1/6th scale model that they made. Its cost was certainly in the bill and the Estate ought to have it. It might be very useful. It would be reasonable to ask to have it simply as a memento.[293]

Prior to this letter, there had obviously been some conversations that have not been recorded because Wilkins replied, on 29th December, stating that he had already been trying to obtain backing from Shell as opposed to Wakefield:

> Thank you very much for your letter which was extremely interesting — I have not lost interest and have gone a long way towards our goal but of course with Shell. I am afraid I did not realise the importance of George Eyston and I did not wish to spend too much money. I went to Shell with a broad outline of a scheme and they are very interested indeed and will I am quite certain play ball with us in a big way if you agree.
>
> Shell have been so good to Leslie Johnson and myself that it would be really impossible to go to Wakefields and if you feel that this would spoil our efforts Reid please let me know as soon as possible.
>
> I have of course left a big loophole with Shell in case we have troubles over engines etc, but otherwise I have told them if we can arrange certain matters such as Saunders-Roe or whoever you decide and Rolls, and most important of all from every point of view Reid Railton, I require £20,000 and I am sure they will agree.
>
> I do hope Reid you will be interested and that by going to Shell I shall not have sabotaged the effort.
>
> I have had one hell of a time over John's affairs, Vicki quite useless and helpless, and bloody difficult — Wilkins losing his temper with Gerard [Cobb's brother] and Vicki, really Reid how John must be laughing the old so and so.

A POINT TO PROVE

A DISTINGUISHED HOTEL
The Whittier
Burns Drive · Detroit 14, · Michigan

December 14th. 1952.

Dear Dick,

How goes the Big Executor? I should love to hear all about it, but I expect I shall have to wait until I see you. I have been doing a lot of thinking about the Crusader episode, and think that I can now see the thing in true perspective, though nothing can excuse the shocking bad judgement that led up to it.

I have also been thinking a lot about possible future activities in that line, and I have come to some fairly definite conclusions as to how to proceed if opportunity should offer. I must confess that I should quite like to prove to the world that the idea was all right, provided we had executed it a bit better.

Are you still interested? I shouldnt be in the least surprised to learn that you have come to your senses, but, in case you have not, I will outline a plan of campaign that I have in mind.

In the first place I think the whole thing should hinge on getting Wakefields' cooperation, and by Wakefields I really mean George Eyston. If we could get him to give us the same whole-hearted support that he gave John, and have him in from the very start (instead of only at the end), then I think I can see a way of handling the job properly. You probably dont realise what a tower of strength he was to us last summer, and how admirably he compensates for my own shortcomings. You would be far better off with Wakefields and George, than with-say- Shell and twice the money.

The programme might then go something like this:-

(1)   I should prepare drawings of the general design over here (and at my own expense).

(2)   Simultaneously I should approach Saunders - Roe through certain private channels, to see if they would be interested.

(3)   I should come over, say in March, and, together with George, try and get an engine out of Rolls Royce, or failing them out of DeHavillands.

Some two and a half months after the accident, Reid Railton wrote this letter to Richard 'Dick' Wilkins, a mutual friend of his and of John Cobb, and Executor of Cobb's will. Railton has obviously not accepted the errors made in executing the design of *Crusader*, and brings up the subject of a new craft. *Railton/Wilkins archives*

> Markswood
> Bishop's Stortford
> Herts
>
> Dec 29th
>
> My dear Reid,
>
> Thank you very much for your letter which was extremely interesting. I have not lost interest and have gone a long way towards our goal but of course with Shell I am afraid I did not realise the importance of George Eyston and as I did not wish to spend too much money I went to Shell with a broad outline of a scheme & they are very interested indeed & will I am quite certain play ball with us in a big way if you agree.
>
> Shell have been so good to Leslie Johnson & myself that it would be really impossible to go to Wakefield and if you feel that this would spoil our efforts Reid please let me know as soon as possible.
>
> I have of course left a big loophole with Shell in case we have troubles over engines etc, but otherwise I have told them if we can arrange certain matters such as Saunders-Roe or whoever you decide, & Rolls, & most important of all from every point of view Reid Railton I require £20000 & I am sure they will agree.
>
> I do hope Reid you will be interested & that by going to Shell I shall not have sabotaged the effort.
>
> I have had one hell of a time over John's affairs. Vicki quite useless & helpless, & bloody difficult – Weekend losing his temper with Gerard & Vicki, really Reid how dear old John must be laughing the old so & so.
>
> John left £90000 of his own and another £30000 which will come to Vicki when Mrs Cobb dies so she is bloody lucky I think.
>
> I paid Vospers all the insurance money so they were paid in full & as I told Du Cane he can very largely

In Wilkins's reply to Railton, it is obvious the seed has been sown. Not only are there possible constructors of a new '*Crusader*' but sponsors too. It is interesting that George Eyston was considering buying Cobb's *Napier-Railton*. Wilkins/Railton archives

John left £90,000 of his own, another £30,000 which will come to Vicki when Mrs Cobb dies so she is bloody lucky I think.

I paid Vosper all the insurance money so they were paid in full and as I told Du Cane he can very largely thank you for that and I must say he is very grateful to both of us.

I will write and ask for the model and I am sure I will get it.

Do please write as soon as you can, telling me what you think of all this and I will then do whatever you think best.

Do you think anybody in the USA would buy John's Car? George Eyston asked for the first refusal which I gave him but have so far heard nothing. My mother sends you and Mrs Railton her best love and of course so do I, and don't be too long before we see you.

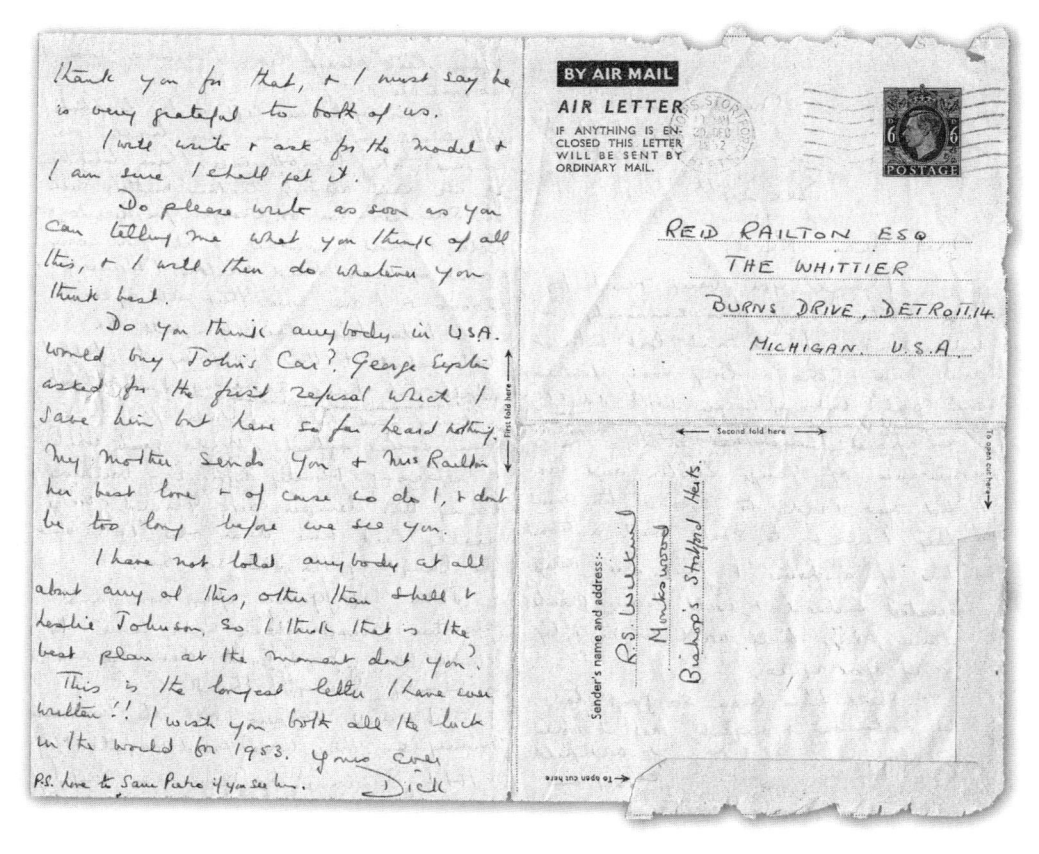

I have not told anybody at all about any of this other than Shell and Leslie Johnson so I think that is the best plan at the moment don't you?

This is the longest letter I have ever written!![294]

Vosper was indeed fortunate, having been paid in full by Cobb, and again by the insurance company for a vessel it did not own!

Not long after this, George Eyston — and therefore C.C. Wakefield and its Castrol brand — became added to the equation, and letters then began to flow freely between the three parties, all eager to get a new boat in the water within two years.

By December 1953, fairly firm plans were in place and a model was being built. Via Stanley Hooker, a jet-engine specialist at the Bristol Aeroplane Company whose earlier jet experience had been with Rolls-Royce, there was tentative agreement to 'borrow' a Rolls-Royce Avon RA3 Mk 101 axial-flow jet of 6,500–7,350lb thrust.

It was becoming clear, however, that there would be another challenger —

CRUSADER

Donald Campbell. On 10th December 1953, Eyston wrote to Railton with early details of Campbell's project:

> I understand that Donald has the services of [the] brother of a draughtsman who did the Hanning-Lee boat. I saw this man at the time and was impressed by the way he was tackling the job. It is presumed that the Campbell's man can refer to the brother who did the former work. [Leo] Villa will be in charge of those employed on the construction work.
>
> I gather that the detail design will not differ from the prototype models at Saunders-Roe. The sponsons will be forward on this first design for 200–240 mph, after which it is contemplated to revert to sponsons aft. Two engines are available, both Metro-Vick (?) as probably explained to you — thrust 3000 lbs (?).[295]

The brothers mentioned by Eyston were Ken and Lewis Norris. Railton responded to this news by saying he had not advised Campbell at all, but had been approached by Norris Bros about analysis information from *Crusader*'s accident. Railton also questioned whether C.C. Wakefield (Castrol) would be in a position to contribute financially to two projects, both British, aiming at the same record. By 16th January 1954, Eyston had the answer for him: "Donald Campbell has been after the money and is prepared to give technical details of his boat next month, February, when a decision will have to be made whether Wakefields go in for it. It has been left for me to decide!"[296]

Two months later, Wakefield decided to support Campbell's new project. However, this did not stop the Railton/Eyston project from progressing to basic tank and wind-tunnel model tests for their proposed attempt in the future, with an unnamed pilot.

Figures from these tests were extremely encouraging, to the point where Railton, at his own expense, began to produce the necessary design drawings to start doing stress calculations. In the meantime, his exchanges with Ken Norris had become more frequent and technically demanding — a challenge that Railton still enjoyed. It was only when working with the Norris brothers on more detailed analysis into the *Crusader* accident, however, that Railton began to feel that perhaps the time had come to pass on the torch rather than proceed further with his own design.

A POINT TO PROVE

George Eyston joins the 'conversation' and mentions (but not by name) Ken Norris's involvement with the *White Hawk* project and the Hanning-Lees. Eyston was in an awkward situation, as sponsorship manager for Castrol, as he wanted Railton to prove a point, but was being badgered by Donald Campbell for the same funds. *Eyston/Railton archives*

> 46, Grosvenor Street,
> London, W.1.
>
> Dec 10th/53
>
> Confidential
>
> Donald Campbell Boat.
>
> My dear Reid,
>
> I understand that Donald has the services of brother of a draughtsman who did the Hanning-Lee boat — I saw this man at the time & was impressed by the way he was tackling the job. It is presumed that the Campbells man can refer to the brother who did the former work. Villa will be in charge of those employed on the constructional work.
>
> I gather that the detailed design will not differ from the prototype models at Saunders Roe. The sponsons will be forward on this finish design for 200-240 m.p.h.
>
> After which it is contemplated to reach to sponsons aft. 2 engines are available, both Metro-Vic (?) as probably explained to you — Thrust 3000 lb (?). It is contemplated to raise money amongst say 15 supporters & the job would cost £9000 to build — ready August 1954.
>
> I expect to know something more of details in the near future but this is just to let you know the rough scheme as it stands at present. I believe it is intended to run at Coniston.
>
> I am not in a position to comment on this matter as I have only received the above second-hand.
>
> Yours George

CRUSADER

He was, he confessed to Leo Villa, "quite tired out by it all".[297]

Ken Norris said he learned much at Railton's side: "Reid was just a normal chap, and we had many happy chats about anything and everything, but when it came to the design work we did, he just came alive, he was hard to keep up with, his slide rule was a blur. I still have one of the ones he gave me. He went through every frame of the *Crusader* crash, and while the lights were out, it was an engineering problem, but once they came on, and the projector switched off, you could see the loss of his friend had troubled him deeply. If Reid told me something I had done looked right, then I could not have been happier."[298]

And with that, Reid Railton passed the torch to Ken Norris. As for *Crusader*, he knew that he had been right about its concept, that poor execution had undoubtedly led to a catastrophic structural failure, and that actually there was little for him to prove.

John Perreit's first paid job had been as a young draughtsman at Thomson & Taylor and working with Railton had a lasting effect on him:

> The climax of Railton's career, as far as cars are concerned, was Cobb's land speed record car, an amazing tour de force of design. Neither Cobb's nor Campbell's land speed record cars ever gave their drivers a moment's worry on their mechanical side, which fact is a tribute to the genius, it was nothing less, of Railton. No one could wish for a better man to work for, he never 'bawled you out' for an error and allowed even I, his most youthful assistant, to have full scope for working my ideas into his schemes. He never lost his cool even when arguing some design point with that most volatile of engineers, Ken Taylor.
>
> Though his work was often hampered by the need to meet a tight budget as there was little sponsors' cash to be had in those days, so his designs needed to be, and were, right at the first go.
>
> Unfortunately, he always suffered from bad health and at the height of his career was obliged to emigrate to California where the more favourable climate enabled him to continue to work, but when in England, he would always visit, and we saw, the fire never left him.[299]

One can only wonder what the fertile mind of Reid Railton would have produced with modern methods such as Computer Aided Design (CAD)

A POINT TO PROVE

Next in line: in her original form, Donald Campbell's *Bluebird K7* arrives at Ullswater in 1955. Following Reid Railton's advice, Ken and Lewis Norris used an incredibly strong metal spaceframe to house both pilot and engine. The sponsons were also mounted on box-section metal beams and the whole craft was of all-metal construction to absorb the loads thought to have been experienced by *Crusader* before the crash. *Getty Images/Mirrorpix*

and Computational Fluid Dynamics (CFD), and what *Crusader* could have achieved if built from high-strength modern materials such as carbon-fibre and Kevlar. Maybe therein lay the problem: Railton's mind was so far ahead of his time that the world needed years to catch up with him.

So touched were local people at Loch Ness that the Glen Urquhart Community Association set up a collection to build a permanent memorial to John Cobb, and schoolchildren even donated pocket money. A stone cairn was built alongside the A82, overlooking the loch at roughly the point where Cobb met his end, and on the first anniversary of his death it was dedicated to his memory in a service that was broadcast on both radio and television. A plaque on the cairn designed by George Bain, a 'Master of Celtic Art', was unveiled by Eileen Holloway, one of Cobb's sisters. Surrounded by a

CRUSADER

border of Celtic Knotwork symbolising eternity, the plaque features an image of *Crusader* at speed and these words:

On the waters of Loch Ness
John Cobb
having travelled at
206 miles per hour
in an attempt to gain the
World's Water
Speed Record
lost his life
in this day
Sept 29th 1952

This memorial is erected as a tribute
To the memory of a gallant gentleman
by the people of Glen Urquhart

Urram do'n Treun, Agus do'n Iriosal
["*Honour to the Brave, And to the Humble*"]

On 27th March 1953, John Cobb was posthumously awarded the Queen's Commendation for Brave Conduct. The official statement read:

John Rhodes Cobb (deceased)
Racing Motorist. For services in attempting to break the world's water speed record and in research into high speed on water, in the course of which he lost his life
<u>Queen's Commendation for Brave Conduct</u>
This Day 27th March 1953

A POINT TO PROVE

A memorial cairn was built alongside the A82 overlooking the spot on Loch Ness where John Cobb lost his life. Funded mainly by those living locally who had taken Cobb as one of their own, it was unveiled by Cobb's sister Eileen, and the ceremony was broadcast on the BBC. *Scottish Press via Gordon Menzies*

POSTSCRIPT

All books, all stories, come to an end. The purpose of this book was always to be factual, to correct past inaccuracies, to present new information and to endeavour to be true to those who lived the events. Having said that, the very nature of this story's conclusion — disaster and tragedy — means that there is inevitably a degree of conjecture over how the analysis led to the conclusions reached. On this occasion, serendipity has also come to the rescue, and history has come to life. The voices of the past have returned to tell us.

During research for this book I became aware of a model that was claimed to be 'something to do with *Crusader*'. I made contact with the owner, Nick Mouat, who filled me in on the background as far as he knew it. When I saw the box that contained the model, the first thing that caught my eye was the handwriting on it — 'Property of G.E.T. Eyston'. I knew that George, whom I met late in his life, had been an Olympic-standard yachtsman and loved sailing — but this model had nothing to do with sailing. The model bore strong similarities to Reid Railton's early concepts for *Crusader* and included a multi-step front similar to the arrangement on the Railton/Van Patten models — but those were designed before Eyston had become part of the team and therefore this newly discovered model was not an early one.

In amongst the letters between Railton and Eyston after the accident, however, I had read that Railton had not only designed a hull form to "show the world there wasn't much wrong with the idea", but that the two men had also begun the search for financial backing and had had a test model made by the National Physical Laboratory. Further research showed that Railton had felt that the flat areas under *Crusader* were too large, so on the new version he had drawn up a more curved, pinched-waist hull, to improve the transition from slow speed to planing. The model had this very feature, plus several others. What Nick owned was *Crusader*'s planned successor. For me, this was useful information, part of the story — but there it might have ended.

During a conversation with Richard Noble, a man for whom speed and the

POSTSCRIPT

The model of the Railton/Eyston successor as found, in its original box, labelled 'Property of G.E.T. Eyston'. *Courtesy of Nick Mouat*

life of John Cobb are food and drink, I mentioned the model and that I had considered buying it, but wondered quite what I would do with it. Richard, as always, was three or four steps ahead, and I brokered a deal that saw a very surprised Nick open his front door to the one-time land speed record holder, who had gone to collect his new model!

Richard then showed the model to *Thrust SSC* designer Ron Ayers, a man who knows a thing or two about vehicles made to go fast, and he became enthralled by some of the model's features. This persuaded Richard that the concept had to be put to the test. To that end, the original model was put through a 3D scanning process carried out by the University of Nottingham Manufacturing Metrology Team and its research partner, Addqual, after which construction of a working model began. Now there arose one of those strange connections that seem to follow record breaking and motorsport: construction of the working model has been done by Len Newton, son-in-law of John 'Lofty' Bennetts, the De Havilland engine technician at Loch Ness during the fateful attempt. At the time of writing, installation of a hand-built scale gas turbine is well in hand.

That Reid Railton's thinking of over half a century ago can still excite and enthral those at the forefront of speed is quite some testament.

The discovery of the model was proof that Railton felt he had been let down, in both the failure to follow his design philosophy and shortcomings in the quality of construction, and thought there was unfinished business. From

CRUSADER

my point of view, what I really wanted was proof that my own analysis of the accident was correct — something most felt to be impossible.

Then, out of the blue, came an email from Craig Wallace of Kongsberg Maritime. Craig, who lives in Scotland and specialises in locating wrecked ships and aeroplanes for governments, has a long-held fascination with John Cobb and *Crusader*. When the time came to test some new equipment, he decided to look for *Crusader*.

> In 2012 working with the Loch Ness Project, Kongsberg Maritime were evaluating the performance of a shallow water Autonomous Underwater Vehicle (AUV), when Adrian Shine of the Loch Ness Project, raised the question of whether we could do some of the testing over the presumed wreck site. Whilst an early attempt was conducted, the difficult terrain and depth meant our efforts were fruitless, but the impression had been made.
>
> From previous aircraft searches I understood quite well that with jet aircraft after a water impact typically the engine is the only piece to remain intact. Indeed, on some aircraft searches I've conducted the debris field can spread hundreds of metres with small difficult-to-detect objects and the 1–2 metre 'lump' of engine is the best chance of locating a distinct target. Given the speed and high deceleration forces with *Crusader*'s crash coupled with the small debris collected and subsequently burned at the time of the accident, it was reasonable to assume *Crusader*'s remains would be similar to an aircraft accident field.

I had made contact with Adrian Shine in the early 1990s and at that time he asked if I had a reasonable idea of where *Crusader*'s engine might be lying. After I sent him copies of the accident investigation maps and the engine blueprint, we stayed in contact sporadically over the years, with Adrian kindly keeping me informed of progress and sending me copies of photographs of small pieces of wreckage. This made me wonder if in fact my analysis was correct, that the hull had not blown apart, releasing the engine, but had ruptured in some way — and that one large piece of wreckage might be further back up the course than had previously been imagined. For Craig Wallace, it was the start of years of searching:

POSTSCRIPT

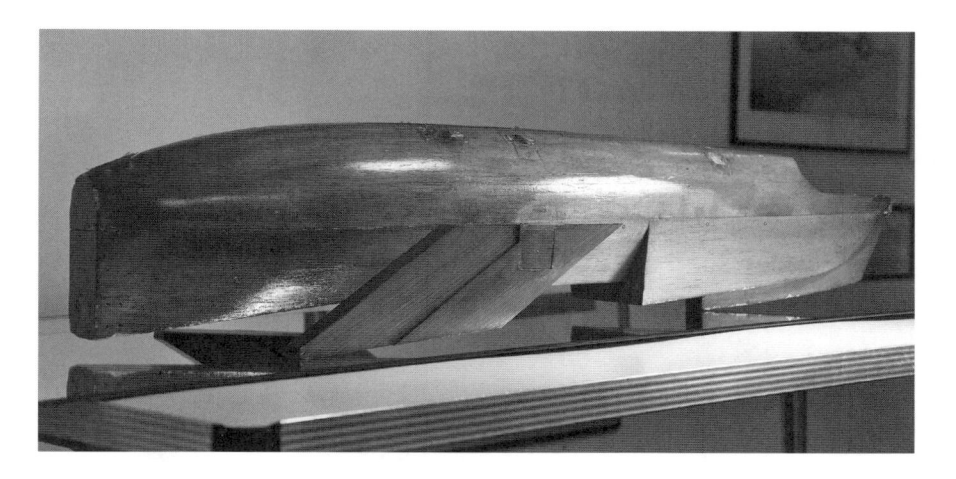

Views of how *Crusader*'s successor might have looked on water. Note the Railton/Van Patten '1½' step and, at the rear, dagger-like sponsons and a rounded underside, to accelerate breakaway. *Courtesy of Richard Noble*

Learning the *Crusader* story through Adrian, we understood some of the wreckage had been filmed by a drop camera [a camera suspended by a very long cable over the side of the vessel] in addition to an ROV [Remotely Operated Vehicle] but in the 220-metre deep loch an accurate position sub-sea was at best a guess from where the vessel was at the time of imaging the debris (within a 400-metre box, though this was in 2005 and subsequently shown to be out of this area). This coupled with never finding the lost jet engine (assumed to be the only piece intact) raised something of a challenge.

CRUSADER

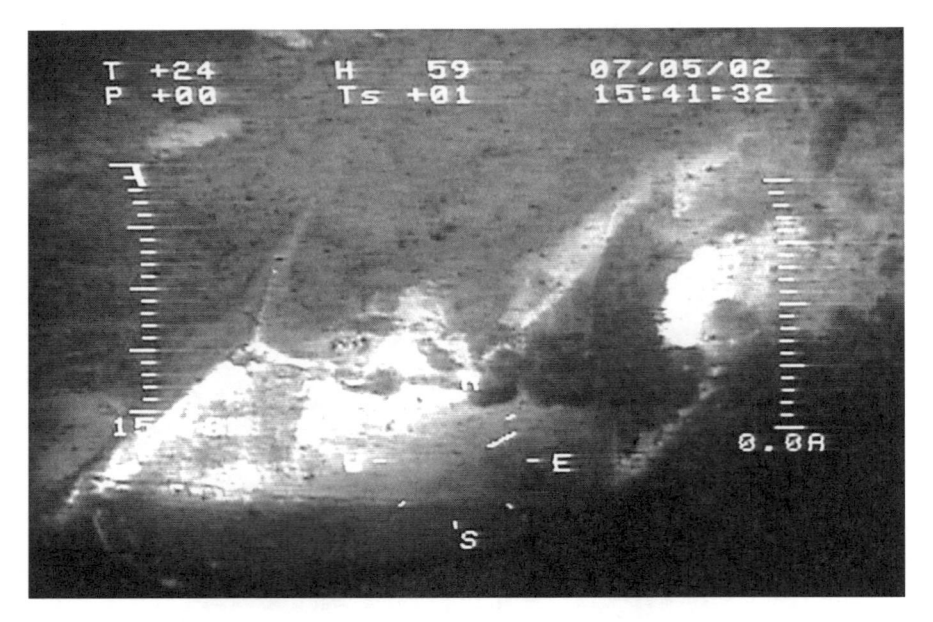

This is a small fragment of wreckage located by Adrian Shine and the Loch Ness Project using a drop camera in 2002. This confirmed to some that there was actually very little wreckage to find. *Courtesy of Adrian Shine and the Loch Ness Project*

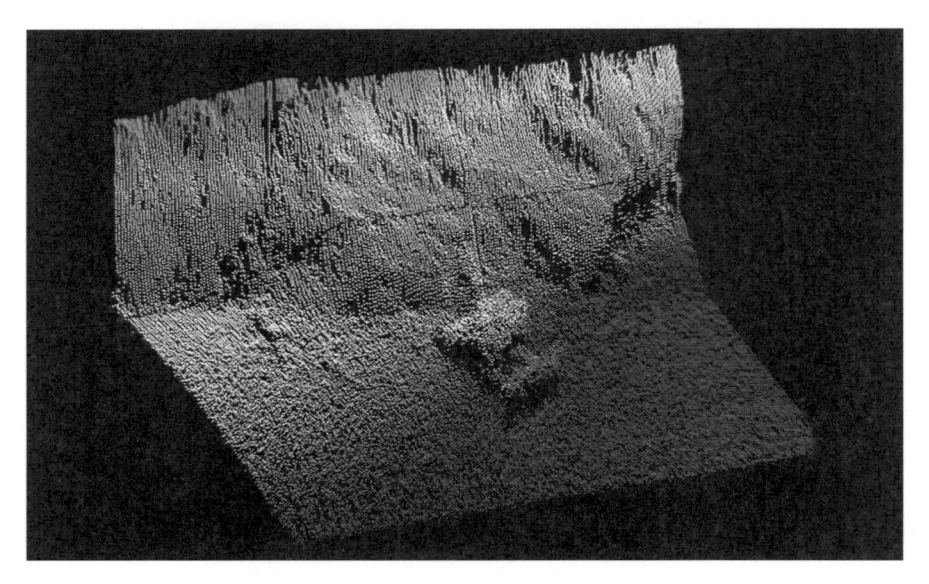

The image that appeared on screen to Craig Wallace of Kongsberg Maritime after the autonomous 'Munin' search vehicle had finally discovered *Crusader*'s resting place. With this one image the author's analysis was confirmed. *Kongsberg Maritime and Craig Wallace*

POSTSCRIPT

When Craig first contacted me, I had only recently finished my analysis and, as such, my assertion that *Crusader* had not blown apart was not known. Craig's theory was that if he could locate the tiny pieces of *Crusader* that were left, save the engine, his new equipment would have proved itself.

The joint forces of Craig Wallace, Kongsberg Maritime and Adrian Shine were bolstered by Gordon Menzies, who had been at his father Alec's side during *Crusader*'s runs on Loch Ness. The search was on — but it was not without its difficulties as Craig explained:

> There are two main components to conducting an underwater search, you need an instrument on a platform to map the seafloor and you need to know where you/the platform is. Traditionally this has been done by towed bodies, platforms dragged behind a surface vessel coupled to instruments such as cameras or sonars which map the area.
>
> Whilst cameras can be used, the range of light in water can be as low as 10 metres and given the high peat content of Loch Ness the water visibility is a few metres at best, hence the field of view is very small in any image, perhaps just 3 metres wide. As a result sonars with the ability to detect even 10cm targets and scan hundreds of metres of sea bed in a single pass become the only feasible instrument in identifying targets. Whilst this now seems a fairly simple task with an instrument such as this, the geological form of Loch Ness fights the sonar. Its steep banks give very strong reflections obscuring the sonar data and making interpretation very difficult. If one can achieve perfect navigation the sonar is improved and you can fly closer to the sea/loch bed and therein lies the second problem. The towed platforms generally have to fly high to avoid the steep terrain the trade-off being lower resolution, which is then compounded by the limited knowledge of the platform's actual geographical location.
>
> Whilst on the surface we all now have a GPS solution on our mobile telephones, but the absorption of electromagnetic waves in water means that no GPS signal can be detected even 10cm into the water. You can estimate the cable length and understanding the depth of the instrument gets you close, but difficulty in towing means you cannot get too close to the sea or loch bed, many a towed body has been lost to the contact with the mud. The solution is to use acoustics or sonar to position the vehicle or platform aiding the operator for turning and

avoiding fast-changing topography. This technology has been available for several decades but traditionally required large vessels with large support crews, something not possible within Loch Ness. Over the last few years Kongsberg Maritime have produced smaller solutions which can be mounted on small vessels and presents a viable solution to position a vehicle within the loch. ROVs with typical maximum speed at 1m/s are generally unstable in flight and as such not the most efficient solution and whilst the real-time operator control is ideal once a target is found and being evaluated you still need to find the target.

This is where the AUV (Autonomous Unmanned Vehicle) comes into play as it is not connected to the surface by a control line; it is an independent vehicle, a robot working on its own allowing it to get closer to the loch or sea bed and in the right circumstances achieves higher resolution than traditional methods but without user interaction and risk.

It was 2016 before we returned with a new vehicle, this time the MUNIN, a miniaturised sibling of the larger HUGIN vehicle, both named after Ravens of Norse Mythology. This was a brand new AUV with 600-metre depth rating and the most advanced sonars available. This vehicle, aimed as a rental solution in the commercial market, required extensive training for the operators and again Loch Ness presented itself as an ideal location for me to push the 'training' programme. The weather whilst inclement is a far more stable solution than the North Sea and gives a far larger weather window for training hence removing risk to the programme and given the fact that the loch is an enclosed water body there is also less chance I would lose the vehicle. Despite two missions on the general crash site the engine again proved elusive and although somewhat disappointed, we did succeed in locating the Loch Ness Monster, well the movie prop version which had been lost during the filming of a 1969 Sherlock Holmes film, so not all was wasted.

For the next two years I spent my spare time researching my old data and looking at options for a privately funded mission to try and return to the loch. Finally, in late 2018 a UK-based production company filming documentaries for the *National Geographic* channel asked if we could return to Loch Ness with MUNIN and whether there was anything new or of interest. Without promising too much I suggested

POSTSCRIPT

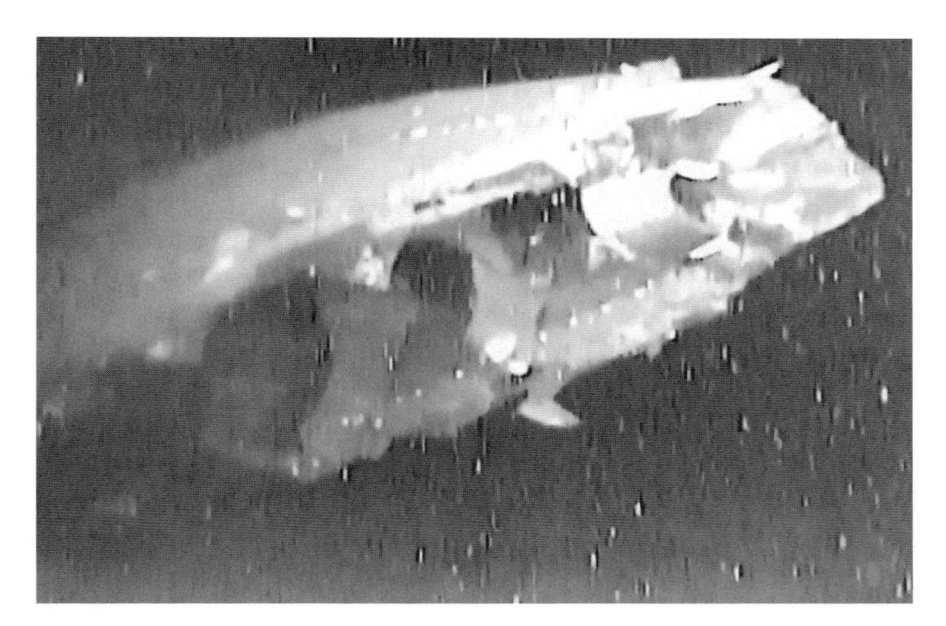

The starboard sponson captured by the remote camera vehicle. At the very front (left of shot) the mounting ring for the accelerometer can be seen. *Craig Wallace/ Kongsberg Maritime*

Looking into the engine bay from where the cockpit would have been. It is clear the hull has snapped where the forward ring bulkhead was installed. Had Reid Railton's three ring bulkheads been fitted, rather than two, perhaps the cockpit would have stayed intact. *Craig Wallace/Kongsberg Maritime*

CRUSADER

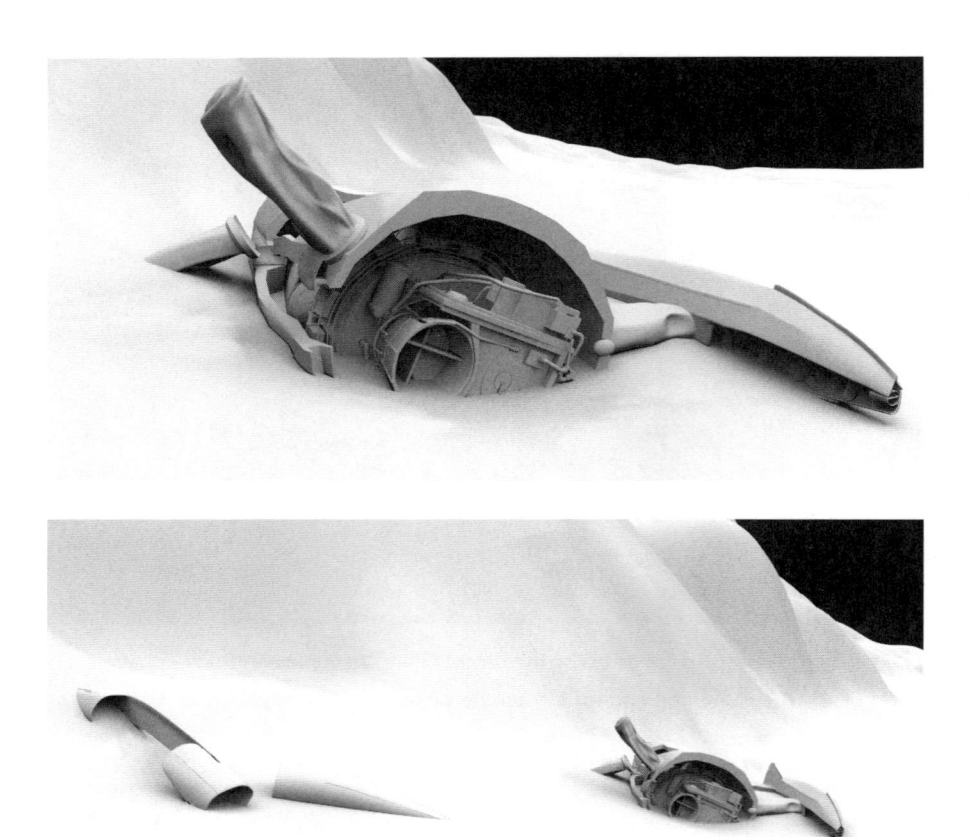

Computer renditions of how *Crusader* lies today at the bottom of Loch Ness — upside down and separated at the exact spot of the forward alloy ring bulkhead. Had the third alloy ring bulkhead been included, nearest the bow, the cockpit would have remained intact. *Courtesy of Kongsberg Maritime, Craig Wallace and Mick Hill*

another search for Cobb's *Crusader*, whilst using known existing shipwrecks as material to justify the risk, and after some meetings in London an agreement was made for a March 2019 expedition.

In all honesty there was little change in hardware with this second visit, but we had gathered experience of the MUNIN's abilities, finding aircraft and shipwrecks all over the world. Our trust in the equipment, coupled to improved understanding of the loch bed around the crash site, meant we could fly lower and achieve better resolution than ever before. Running one mission close to the steep cliff edge, the data

POSTSCRIPT

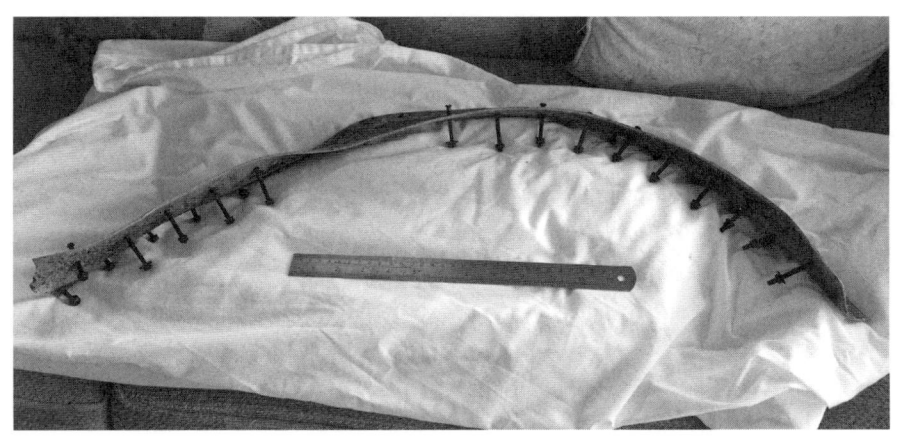

Mementoes of *Crusader*: the aluminium cockpit surround (taken off one of the pieces of floating wreckage) and one of the triangular plates cut from the forward planing surface brace brackets, bearing the signatures of Hughie Jones and Geoff Brading.*Courtesy of Gordon Menzies*

was downloaded for processing and analysis, and then, as I sat in a Drumnadrochit coffee shop studying the multibeam sonar data on a laptop, *Crusader* presented herself. Six years after our first attempt and 67 years after the accident, we saw the colored data of what was almost certainly *Crusader*. Lying upside down in the soft sediment, not a jet engine but the complete hull aft of the cockpit including the starboard sponson and port sponson support legs.

Using the sonar data, we were already questioning the accident's nature. We could see some small presumed debris fields around the main crash site which extends only some 50 metres but with the location of the primary wreck only a metre from a cliff edge one wonders if *Crusader* had tumbled down the banks of Loch Ness. It was these questions that had in turn led me to the internet searches that would put me in touch with Steve [Holter] and his book on the accident. Whilst to me the colored images were undoubtedly a conclusion on years of question and frustration, photographic evidence is required to prove to the world, *Crusader* had been found. And so, we began to plan an ROV deployment using our acoustic positioning system to dive and film the sonar targets categorically defining this as the *Crusader*.

Despite huge problems with weather, a team so passionate about *Crusader*, we found ways to get the job done and when we discovered that the recorder hadn't worked we went out and did it again!

When Craig sent me that very image, it was obvious that it was the unmistakable shape of *Crusader*, lying at the bottom of a steep underwater cliff, upside down, missing her starboard sponson and the entire bow, but otherwise complete. In that instant, I knew my analysis was right.

As Craig sent more data, it became apparent that not only had the bow snapped off as I had thought, but also that the section that remained intact was the area between the alloy ring bulkheads, the boat's hull forward of this being completely missing.

My immediate thought was this: if Reid Railton's forward alloy hoop had been built in, the cockpit would still be there...

APPENDIX

Memories of Alec Menzies

A native of Lewiston, near Drumnadrochit, my father James Alexander (Alec) Menzies (1901–88) moved to Glasgow on leaving the local secondary school to serve an apprenticeship as a boilermaker. On completing this he joined the Merchant Navy as a marine engineer. In 1929 he married my mother, a Glasgow lass, Margaret (Peggy) Montgomery, and in 1932, having resigned from the Merchant Navy, he returned to Lewiston with Peggy and their infant daughter, Dorothy.

At Lewiston he and his father Jimmy (the local blacksmith) built 'The Garage', a vehicle-repair premises with an associated filling station. This is where he spent the night smoothing out the larger, cast, Vosper rudder before it was fitted to *Crusader*. It is also where, I believe, the large metal triangles fitted atop the measured mile posts were, in part, fabricated. In 1935, the tenancy of

Taken at a local Friday night dance and captioned 'Neatest Ankles Contest', this photo shows the winning lady in company with Mr and Mrs Cobb and Alec Menzies.
Courtesy of Gordon Menzies

the smallholding at Temple Pier became vacant and Alec successfully applied for it. As tenant he also became Pier Master with the paid responsibility (on commission) to collect pier dues from anyone using the services of the commercial paddle steamers which then plied the Caledonian Canal between Inverness and Fort William.

This service was discontinued in 1935 as the new A82 trunk road between Inverness and Glasgow had come into use. At some point after his arrival at Temple Pier, Alec set up a second business of coal deliveries to the surrounding villages, and up until the middle to late 1940s the pier continued to be used to take bulk deliveries of coal by small coastal steamers. (I remember the last one to come here — the *Nansen* — crewed by Norwegians). The expansion of the railways led to the demise of these vessels, so the pier fell into disrepair through lack of use. My father had by this time (in 1945) bought Temple Pier outright during the sale of the local Seafield Estate.

Our family had by now also increased to seven, four girls and three boys.

He had a deep and abiding interest in local community affairs and served for many years as a District Councillor as well as on the committees of The Highland Games, Community Association and Public Hall, to name but a few. Thus when he was approached by C.C. Wakefield (Castrol Oils) in 1952 about the possibility of Temple Pier becoming the base for John Cobb's attempt on the world water speed record, he was more than delighted to agree to what he saw as a wonderful opportunity for the area.

Gordon Menzies (son)

Memories of John 'Lofty' Bennetts

My Dad, John James Bennetts (1920–92), was known by most as 'Lofty' due to his 6ft 6in stature. Born into a farming family, he spent hours from an early age dismantling anything mechanical to see how it worked. He won a scholarship for Watford Engineering College and, following his first year at Watford, joined De Havilland in 1936 as an apprentice at Stag Lane and later moved to Hatfield where he played a key role as an engineer on the 'H1 Supercharger', later known as the Goblin jet engine.

He became involved in the development and testing of the Ghost engine alongside George 'Guy' Bristow, who was known as one of the best development engineers De Havilland ever had. 'Lofty' and Guy soon became known as the 'DH Duo' and were well respected in the aviation industry for their outstanding knowledge of gas-turbine engines.

APPENDIX

John 'Lofty' Bennetts tends his charge at Loch Ness — the De Havilland Ghost engine. He quickly became part of the *Crusader* 'family'. *Courtesy of Julie Newton*

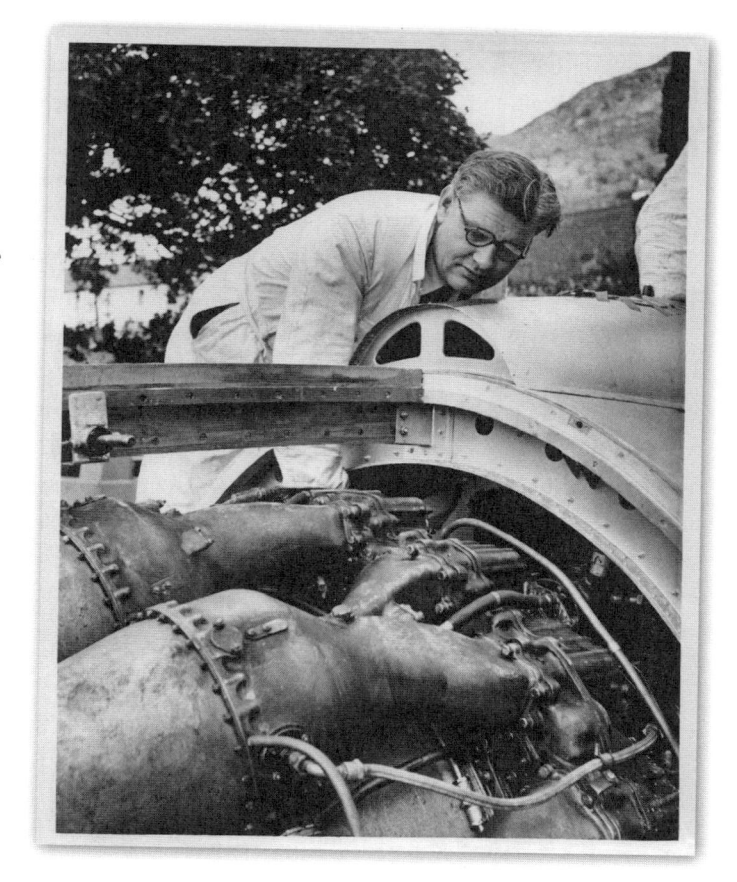

In 1952 he was seconded to support the *Crusader* project. He installed and tested the engine and fuel system at Portchester Harbour, then travelled north to Loch Ness to reinstall and test it again after its removal for transportation.

'Lofty' became a favourite with the locals. Always full of fun and laughter wherever he went, he was welcomed and loved by all. He accompanied locals on shoots, and he and Guy Bristow would often go off rabbit shooting in the heather hills above the Loch.

In one of his letters to my Mum, Sue, dated 25th September, he mentions the pouring rain holding up planned schedules and in his words: "Bris and I have the job of checking up on floating wood in the Loch, trees and other stuff washed down by the floods, Bris is at present jawing to some locals, trying to get some gen on the rise of water and which of the debris floats on the Loch, I'm happy, sitting in the van out of the rain."

Although, 'Lofty' was accustomed to working away from home, this trip was quite a wrench as he had recently married Sue (Boyton-Salts). He wrote romantic letters most nights, always suggesting he wished Sue could join him, his reasons being that he was cold at night, was in a remote location and missed her tremendously — a big softy at heart.

'Lofty' held John Cobb in high esteem and was truly saddened by the accident and his death. He never really spoke about the water speed record attempt, but always spoke about the months and weeks before that awful tragic day in September 1952.

'Lofty' remained with De Havilland until taken over by Bristol Siddeley then Rolls-Royce. From 1956 until retirement in 1982, his working life was devoted to the Royal Navy and at times the Tri-Services overseeing the Vampire, the Venom and the Buccaneer. Moving on to Rotary wing, including the Wessex and the Gnome-engined Sea Kings, he was instrumental in the development of coloured smoke, as now used by the Red Arrows display team. He remained with the Royal Navy visiting many different overseas establishments and onboard many aircraft carriers throughout the world.

He was known and respected throughout the Tri-Services as 'Lofty' or 'Mr Rolls-Royce'.

He was a true gentleman full of character and charisma — once met never forgotten.

My Dad — My Hero.

Julie and Leonard Newton (daughter and son-in-law)

Richard Noble's perspectives

I came across the water speed record when I was six. Father took the family for a drive around Loch Ness in Scotland and we stopped at Temple Pier to look over the wall and gaze at the silver and red *Crusader* on its transport pallet sitting on the jetty. I can still picture that image all these years later and something clicked — the engineering achievement, the sheer audacity of what was being attempted. Then I discovered there was a land speed record and that was so much faster...

Back in 1952, the *Crusader* hit a wave and broke up at over 200mph, killing John Cobb. It wasn't until the publication of the Evro book about Reid Railton that we were able to learn more. The *Crusader* designers, Reid Railton and Peter Du Cane, were trying to predict performance above the existing record of 178mph but they could never have imagined the boat hitting a serious wave at 200mph.

Railton went back to California and, keen to prove the accident was not due

APPENDIX

to a design fault, spent some 18 months designing a new boat. He and George Eyston were not able to progress beyond test models as there was no finance available. The tank model stayed in its box and 65 years later I was able to acquire it. With approval from the Railton family, we started a programme to develop it and learn its secrets. We call it *Crusader 2*.

Much later I was able to inspect Donald Campbell's *Bluebird K7* and was fascinated by the design complexity: it was never just a blue boat with a jet engine — the hydro side of the boat was extremely complicated. Everywhere you looked there was another of Ken Norris's carefully thought-through features.

The water speed record has been held by Ken Warby since 1978 and today's challengers seem to have very similar designs to his original *Spirit of Australia*. This suggests that it's likely that the record will plateau since similar challengers may only be able to achieve similar performance. It's a bit like the airliners of today: it's often difficult to tell the difference between a Boeing and an Airbus — they look the same and have similar performance.

But Railton's *Crusader* was different. It was a tricycle layout, a reverse three-pointer with one planing point at the front. Railton realised that to achieve a significant record advance a very different design was needed. And *Crusader* is believed to have peaked at 220mph.

This set me thinking and understanding what an incredibly difficult technical challenge the water speed record is. There have to be so many different design compromises. The boat has to float safely in displacement mode, accelerate away fast and rise up onto the plane. The design has to plane safely over a wide speed range: it has to have minimum aero drag and excellent aero stability below 200mph when the aero fin will contribute little, and it must not fly so the stability in pitch has to be extreme.

Operating on a short stretch of calm water, the boat has to accelerate quickly and decelerate even more quickly and this is not helped by the planing drag which I understand is minimal — making the boat very quick to accelerate and difficult to slow.

The next record step is likely to be around 350–400mph and that is a mighty challenge requiring jet-fighter-like performance, huge power and violent but dependable acceleration and deceleration. The ultimate drag race. If it ever happens, it is going to be mighty spectacular.

All this confirms what an outstanding job Ken Warby did all those years ago!

Richard Noble

ENDNOTES

Some correspondence is listed in the order of who sent it/who received it. This is because the letters in question exist as both the sender's carbon copy and the recipient's copy; sometimes the latter has hand-written additions that may not be on the carbon copy.

1	Interview with author	40	Unpublished article, Cobb family archive	63	Cobb/Railton correspondence
2	Railton family archive	41	Interview with author	64	Cobb/Railton correspondence
3	Cobb family archive	42	Interview with author	65	Correspondence with author
4	Cobb family archive	43	Various magazines and books including *An Engineer of Sorts* by Peter Du Cane	66	Cobb/Railton correspondence
5	Interview/correspondence with author			67	Cobb/Railton correspondence
6	Interview with author				
7	Railton personal archive				
8	Interview with author	44	Railton/Cobb/Du Cane correspondence	68	Cobb/Railton correspondence
9	Interview with author				
10	Cobb family archive	45	Railton/Cobb/Du Cane correspondence	69	Cobb/Railton correspondence
11	Interview with author				
12	Interview/correspondence with author	46	Railton/Cobb/Du Cane correspondence	70	Cobb/Railton correspondence
13	Interview with author	47	Railton/Cobb/Du Cane correspondence	71	Cobb/Railton correspondence
14	Interview/correspondence with author	48	Railton/Cobb correspondence	72	Cobb/Railton correspondence
15	Via Doug Nye				
16	Interview with author	49	Railton/Du Cane correspondence	73	Cobb/Railton correspondence
17	Interview with author				
18	Correspondence with author	50	Interview with author		
19	Interview with author	51	Railton/Cobb correspondence	74	Cobb/Railton correspondence
20	Castrol archive				
21	Correspondence with author	52	Railton/Cobb correspondence	75	Du Cane/Railton correspondence
22	Interview with author				
23	Interview with author	53	Railton/Du Cane correspondence	76	Railton/Cobb correspondence
24	Interview with author				
25	Rolls-Royce Heritage archive	54	Railton/Cobb correspondence	77	Railton/Cobb correspondence
26	Campbell family archive	55	Cobb family archive	78	Railton/Cobb correspondence
27	Vosper/Hampshire archive	56	Railton/Cobb correspondence		
28	Campbell archive			79	Railton/Cobb correspondence
29	Campbell archive	57	Cobb family archive		
30	Interview with author	58	Cobb/Railton correspondence	80	Railton/Cobb correspondence
31	Interview with author				
32	Railton/Villa archive	59	Cobb/Railton correspondence	81	Railton/Cobb correspondence
33	Interview with author				
34	Railton archive	60	Cobb/Railton correspondence	82	Railton/Cobb correspondence
35	Interview with author				
36	Interview with author	61	Cobb/Railton correspondence	83	Correspondence with author
37	Castrol archive			84	Correspondence with author
38	Interview with author	62	Cobb/Railton correspondence	85	Railton/Cobb correspondence
39	Interview with author				

ENDNOTES

86 Cobb/Railton correspondence
87 Railton/Du Cane correspondence
88 Cobb/Railton correspondence
89 Cobb/Railton correspondence
90 Railton/Du Cane correspondence
91 Railton/Du Cane correspondence
92 Cobb/Railton correspondence
93 Cobb/Railton correspondence
94 Du Cane/Railton correspondence
95 Cobb/Railton correspondence
96 Cobb/Railton correspondence
97 Railton/Cobb correspondence
98 Cobb/Railton correspondence
99 Du Cane/Railton correspondence
100 Railton/Du Cane correspondence
101 Railton/Cobb correspondence
102 Interview with author
103 Cobb/Railton correspondence
104 Railton/Cobb correspondence
105 Cobb/Railton correspondence
106 Du Cane/Vosper/Railton/Cobb correspondence
107 Railton/Cobb correspondence
108 Railton/Du Cane correspondence
109 Railton family archive
110 Railton family archive
111 Du Cane/Railton correspondence
112 Du Cane/Railton correspondence
113 Railton/Cobb correspondence
114 Cobb/Railton correspondence
115 Cobb/Railton correspondence
116 Railton/Cobb correspondence
117 Cobb/Railton correspondence
118 Railton/Cobb correspondence
119 Railton/Cobb correspondence
120 Cobb/Railton correspondence
121 Railton/Cobb correspondence
122 Railton/Cobb correspondence
123 Cobb/Railton correspondence
124 Railton/Cobb correspondence
125 Du Cane/Railton correspondence
126 Cobb family archive
127 Du Cane/Cobb/Railton correspondence
128 Cobb/Railton correspondence
129 Railton/Cobb correspondence
130 Cobb/Railton correspondence
131 Railton/Cobb correspondence
132 Correspondence with author
133 Interview with author
134 Railton/Cobb correspondence
135 Cobb/Railton correspondence
136 Du Cane/Railton correspondence
137 Railton/Cobb correspondence
138 Railton/Du Cane correspondence
139 Interview with author
140 Du Cane/Railton correspondence
141 Railton/Cobb correspondence
142 Du Cane/Railton correspondence
143 Du Cane/Railton correspondence
144 Du Cane/Railton correspondence
145 Railton/Du Cane correspondence
146 Cobb/Railton correspondence
147 Railton/Cobb correspondence
148 Railton/Cobb correspondence
149 Du Cane/Railton correspondence
150 Cobb family archive
151 Cobb/Railton correspondence
152 Article in *Motor Boat and Yachting*, August 1952
153 Cobb/Railton correspondence
154 Railton/Cobb correspondence
155 Railton family archive
156 Du Cane/Railton correspondence
157 Railton/Du Cane correspondence
158 Railton/Du Cane correspondence
159 Du Cane/Railton correspondence
160 Railton/Cobb correspondence
161 Railton family archive
162 Cobb/Railton correspondence
163 Railton/Cobb correspondence
164 Du Cane/Railton correspondence
165 Cobb family archiive
166 Railton/Cobb correspondence
167 Railton/Du Cane correspondence
168 Railton/Cobb correspondence
169 Du Cane/Railton correspondence
170 Du Cane/Railton/Cobb correspondence
171 Cobb/Railton correspondence
172 Railton/Cobb correspondence
173 Cobb/Railton correspondence
174 Railton/Cobb correspondence
175 Correspondence with author
176 Cobb/Railton correspondence

177 Railton/Cobb correspondence
178 Railton/Du Cane correspondence
179 Railton/Cobb correspondence
180 Cobb/Railton correspondence
181 Cobb/Railton correspondence
182 Railton/Cobb correspondence
183 Railton/Du Cane correspondence
184 Railton/Du Cane correspondence
185 Correspondence with author
186 Article in *Motor Boat and Yachting*, August 1952
187 Railton/Cobb correspondence
188 Du Cane/Vosper/Hampshire archive
189 Du Cane archive
190 Vosper/Hampshire archive
191 Vosper/Hampshire archive
192 Correspondence with author
193 Ken Norris archive
194 Vosper/Hampshire archive
195 Cobb/Railton correspondence
196 Vosper/Hampshire archive
197 Vosper/Hampshire archive
198 Cobb family archive
199 Pathé News interview (unused)
200 Alec Menzies article for Scottish Press
201 *Glasgow Herald*
202 Interview with author
203 Interview with author
204 Interview with author
205 Interview with author
206 Interview with author
207 Interview with author
208 Vosper/Hampshire archive
209 Interview with author
210 Correspondence with author
211 Interview with author
212 Interview with author
213 Interview with author
214 Interview with author
215 Vosper/Hampshire archive
216 Interview with author
217 Interview with author
218 Du Cane/Vosper archive
219 Interview with Associated Press
220 Interview with *Motor Boat and Yachting*
221 Article in *Motor Boat and Yachting*, August 1952
222 Interview with author
223 Interview with author
224 Du Cane archive
225 Correspondence with author
226 Interview with author
227 Interview with author
228 Vosper/Hampshire archive
229 Railton/Du Cane correspondence
230 Railton/Vosper/Campbell archive
231 Railton/Du Cane correspondence
232 Interview with author
233 Correspondence with author
234 Correspondence with author
235 Correspondence with author
236 Interview with author
237 *Highland Gazette*
238 Correspondence with author
239 Interview with author
240 Correspondence with author
241 *New York Times*
242 Correspondence with author
243 Correspondence Kevin Desmond/Du Cane
244 Eyston, *Safety Last*
245 *Highland Gazette*
246 Interview with author
247 *Evening Standard*
248 Vosper/Hampshire archive
249 Vosper/Hampshire archive
250 Eyston archive via Denly
251 Interview BBC (BBC Scotland archive)
252 Report to Marine Motoring Association
253 Interview with author
254 Report to Marine Motoring Association
255 Interview with author
256 Vosper/Hampshire archive
257 Interview with author
258 Interview with author
259 Vosper/Hampshire archive
260 *Moriston Matters*, February and April 1979
261 Correspondence with author
262 Correspondence with author
263 Correspondence with author
264 Interview with author
265 Interview with author
266 Correspondence with author
267 Interview with BBC (BBC Scotland Archive)
268 Correspondence Kevin Desmond/Du Cane
269 Du Cane archive
270 Correspondence with author
271 Correspondence with author
272 Interview with author
273 Marine Motoring Association/FIM archive
274 William Maloney SSR Publishing article
275 Interview with author
276 Interview with author
277 Interview with author
278 Interview with author
279 Interview with author
280 Interview with author
281 Interview with author
282 Railton archive
283 Correspondence with author
284 Correspondence with author
285 Du Cane archive
286 Du Cane archive
287 Marine Motoring Association/FIM archive
288 Correspondence with author
289 Hampshire archive
290 Du Cane/Railton archive
291 Interview with author
292 Railton/Wilkins archive
293 Railton/Wilkins archive
294 Wilkins/Railton archive
295 Eyston/Railton archive
296 Eyston/Railton archive
297 Interview with author
298 Interview with author
299 Interview with author

INDEX

Adams Transport 53, 197, 198–199
Admiralty Experiment Works 52, 80, 82, 84, 90, 98, 99, 102, 108, 115, 121, 124, 125, 130, 135, 143, 146, 153, 157
Air Ministry 54
Air Transport Auxiliary 40
Aitken, Max 221
Aluminium alloys
 DTD 213 182
 DTD 610 217, 218
  DTD 610B 79, 152, 160, 183
 Duralumin 46, 50, 150
 RR 53 71
Anning Chadwick & Cobb (fur broker, name until 1929) 32, 38
Anning Chadwick & Kiver (fur broker, name from 1930) 38, 100, 102, 108
Anning, Henry 32
Apel, Adolphe E. 49, 50, 51, 52, 75, 89, 93, 94, 107, 116, 120
Arab (cars) 22–23, 24, 28
 Super Sports 28
Armstrong Whitworth (aircraft) 288
Astraea II (hydroplane) 84–85, 108
Astrid (boat) 206, 240, 241, 242, 245, 247, 267, 277, 279
Austin truck 197, 203
Ayers, Ron 329

Babs (record car) 24–28, 31, 35
BAE Systems 12
Bain, George 326
Barclay, Jack 38
Barr, Angus 235
Batchelor, Denzil 205
BBC (British Broadcasting Corporation) 250, 252, 259, 260–261, 275, 276, 278, 291, 294, 296, 306, 327
 BBC Scotland 259
Beauchamp, Reg 24, 25, 27, 28, 49, 52, 58, 60, 64, 66, 70, 73–74, 78–79, 85, 102, 118, 125, 147, 151, 152, 163–164, 183, 218, 221, 239, 313
Bennetts, John 'Lofty' 201, 210, 212, 229, 267, 269, 329, 340–342
Bennetts, Sue (*née* Boyton-Salts) 341–342
Bentley (cars) 36, 37
Benz (cars) 24
Berthon, Peter 30
Bertram, Oliver 37
Birkin, Henry 'Tim' 36, 37
Bismarck (warship) 124
Black, Sir John 31
Blue Ace (boat) 49, 51, 52
Blue Bird (car, Malcolm Campbell) 27, 29–30, 45, 47, 54, 63
Blue Bird (boat, Malcolm Campbell)
 K3 45, 46–47, 48, 49, 50, 51, 54, 78, 94, 166, 175, 186, 224
 K4 17, 49, 50, 52, 53, 54–62, 73, 78, 80, 81, 90, 93, 94, 95, 98, 101, 102, 117, 129, 130, 132, 140, 143, 170, 175, 176, 177, 181, 186, 224, 237, 288
Bluebird (car, Donald Campbell) 289
 CN7 289, 291
Bluebird (boat, Donald Campbell)
 K7 50, 177, 181, 286–287, 289, 302, 325, 343
Bonneville Salt Flats 29, 44, 45, 63, 66, 68–71, 76, 87, 93, 201, 234, 291
Boothman, Air Marshal John 145
Brading, Geoff 182, 185, 189, 192, 212, 227, 235, 236, 337
Bray, Lieutenant Commander Arthur 101, 102, 233, 234, 235, 240, 241, 244, 246, 248, 249, 250, 262, 263, 264, 270, 273, 279, 280–286, 308
Brinton, Lieutenant G.L. 'Monty' 305
Bristol Aeroplane Company 321
Bristol Siddeley 342
Bristow, George 'Guy' 201, 215, 216, 225, 232, 267, 269, 340, 341
British Aluminium 54, 71, 73, 78, 94, 170, 171, 217
BRM (British Racing Motors) 30–31
Brooklands 11, 21, 22, 23, 24, 25, 27–28, 31, 33–34, 38, 39, 42, 44, 52, 53, 55, 64, 86, 126, 236
 '90mph Short Handicap', 1927 28
 '100mph Short Handicap', 1927 38
 '100mph Long Handicap', 1927 38
 British Empire Trophy race, 1932 36, 37
 Easter Meeting
  1924 23
  1937 39
 Gold Star race, 1927 38
 Outer Circuit 24, 38, 39, 43
 Summer Meeting, 1925 35
 'The Hermitage' 22, 23, 25, 27
 Whitsun Meeting, 1926
Bruce, R.A. 307–308

Campbell, Donald 11, 29, 50, 52, 53, 60–62, 79–80, 81, 87, 94, 95, 96, 97, 98, 101, 170, 181, 202, 287, 288, 289–291, 296, 322, 323, 325, 343
Campbell, Lorne 96, 145–146, 174, 177, 181, 278, 297, 299, 307
Campbell, Malcolm (Sir) 11, 17, 25, 27, 34, 41, 42, 45–49, 51, 52, 53, 54–55, 56–60, 63, 70, 73, 81, 84, 90, 93, 94, 95, 96, 97, 98, 140, 144, 166, 169, 176, 177, 202, 222, 237, 324

CRUSADER

Carpenter, Jean 113
Carstairs, Betty 63
Castrol (oil) 12, 72, 101, 165, 166, 172, 200, 206, 235, 273, 286, 317, 318, 321, 323, 340
C.C. Wakefield Ltd — see 'Castrol'
Chadwick, William 32
Chapman, Colin 222
Christian, R.L. 285
Coandă effect 304, 305, 306, 312
Coandă, Henri 304
Cobb, Ann (née Rhodes) (ancestor) 32
Cobb, Arthur R. (ancestor) 32
Cobb, Eileen Lucy (sister) 32, 33, 38, 42, 326, 327
Cobb, Elizabeth (née Mitchell-Smith) (first wife) 40–41, 67
Cobb, Florence (née Goad) (mother) 32, 40–41
Cobb, Frederick E. (great uncle) 32
Cobb, Gerard Rhodes (brother) 32, 34, 40, 320
Cobb, John 6, 8, 11, 12, 17, 19, 31–44, 49, 58, 60, 63–64, 66–72, 73, 74, 76, 77, 78, 79, 80, 81, 82, 83, 84, 86, 87, 88, 89, 90, 92–128, 129, 133–136, 137–149, 152, 154–156, 158–160, 162, 165–176, 180, 182, 189, 192, 193, 194, 197, 199, 200, 201, 202, 203, 204, 205–207, 210–211, 214, 215, 216, 217, 221, 222, 225, 227, 228, 230, 231, 232, 233, 234–235, 236–237, 238, 239, 240–250, 252, 254, 258, 259, 260–261, 262, 263, 264, 266, 267, 269, 270, 272, 273, 274–276, 278, 279, 280, 281, 282, 284, 285, 286, 288, 289, 290, 291–293, 294, 296, 297, 300, 302, 307, 308, 313, 315, 316, 317, 318, 319, 324, 325–327, 328, 330, 334, 339, 340, 342
Cobb, John Hawtyn (nephew) 43–44
Cobb, Lydia (née Davies) (grandmother) 32
Cobb, Olive (sister) 32, 40

Cobb, Rhodes (grandfather) 32
Cobb, Rhodes Hawtyn (father) 32
Cobb, Rhodes Stanley (brother) 32, 33
Cobb, Sarah (née Hawtyn) (ancestor) 32
Cobb, Thomas (ancestor) 32
Cobb, Timothy (ancestor) 32
Cobb, Timothy Rhodes (ancestor) 32
Cobb, Vera Victoria 'Vicki' (née Siddle) (second wife) 41–42, 165, 205, 237, 242, 245, 250, 251, 252, 269, 270–271, 272, 273, 274, 284, 285, 286, 318–319, 339
Cole, Harry 195, 198, 201, 204, 208, 209, 210, 212, 232, 242, 248, 285, 302
Collins, Michael 22
Coniston Water (Lake District) 55, 56, 57, 58, 59, 60, 61, 62, 156, 166, 169, 170, 176, 289
Cook, Humphrey 30
Cooper, Fred 45, 46, 48, 49, 50, 52, 53, 224
Cooper, Jack 'Pop' 51
Cope, Ted 205, 232, 241, 242, 244, 245, 247, 250, 263, 271, 272
Corlett, Dr Ewan 54, 62, 71, 73, 74, 78, 94, 97, 102, 118, 152, 160, 164, 167, 169, 170, 171, 183, 184, 185, 186, 217, 218, 221, 226, 227, 234, 288
Crewe, Peter 84, 97, 159, 183
Cronk, Basil 201, 204, 205, 207, 210, 212, 214, 217, 225, 233, 253, 279
Cronk, Dennis 214, 217
Cronk, Joyce 253–256, 264, 269, 270, 271, 273
Crouch, George W. 51

Daily Express 221
David Brown & Sons 66
Davis, S.C.H. 'Sammy' 41, 42
Daytona Beach, Florida, USA 27, 29
De Dion (cars) 33
De Havilland (aircraft and engines) 56, 72, 73, 81, 84, 94, 97, 104, 107, 114, 182, 201, 203, 207, 215, 267, 317, 329, 340, 341, 342

Comet 182
Ghost jet engine 84, 94, 95, 107, 117, 147, 163, 182, 188, 201, 203, 212, 228, 244, 303, 341
48 Mk I 182
Goblin jet engine 56, 84, 94, 95, 107, 117, 182, 340
Vampire 342
Sea Venom 182, 342
Venom 182, 342
Delage (cars)
V12 36–37
Denly, Bert 201–202, 204, 205, 207, 210, 211, 214, 215, 217, 219, 221, 227, 234, 239, 242, 251–252, 266–267, 269, 270, 274, 276, 279, 280, 297, 308
Desmond, Kevin 309
Dixon, Freddie 30, 87
Don, Kaye 166, 170
Du Cane, Charles 214, 216, 232
Du Cane, Peter 8, 12, 17, 47, 49, 52, 54–55, 56, 57, 58, 59, 73, 79–80, 82–83, 84, 86, 89–92, 93, 94, 95, 96, 97–99, 101, 103, 104, 107, 108, 109, 110, 112–126, 128–129, 130, 132–136, 137–163, 165, 166–177, 180–181, 183, 184, 186, 189, 192, 193, 194–197, 199, 202, 205, 207, 210, 211, 213, 214, 215–217, 218, 221, 222, 223, 224–225, 226, 227, 228–230, 232, 233, 234–235, 239, 245, 250, 254, 257–258, 264–266, 267, 269, 270, 271, 272, 273, 274, 275, 276, 278, 279, 280, 282, 286, 297, 299, 300, 302, 303, 304, 308, 309, 320, 342
Duff, John 35
Dunlop (tyres) 27, 44, 93

Ebblewhite, A.V. 'Ebby' 36
Eddy Marine Corporation 51
Eddy AquaFlow 51
Eldridge, Ernest 34–35, 36
Enfield-Allday (cars) 23
ENV (rear axles) 23
ERA (English Racing Automobiles) 11, 30
Estelle (motorboat) 64

348

INDEX

Eyston, George 12, 36, 37, 63, 64, 65, 66, 68, 94, 96, 98, 100, 101, 102, 104, 107, 108, 146, 165, 166, 167, 171, 200, 201, 202, 205, 207, 210, 215, 217, 225, 227, 234, 236, 237, 242, 246, 251, 252, 254, 267, 270, 274, 275, 276, 286, 287, 317, 318, 320–321, 322, 323, 328, 329, 343

Fairey Aviation Company 124–125, 135, 145, 155, 157, 167, 168, 169, 170, 171, 172, 193
   Swordfish 124
Falkland Islands Company 32, 275–276
Ferguson tractor 201
FIAT *Tipo* S.61 35
Fink, Professor P.T. 'Tom' 181, 291, 296, 302
Ford (engines)
   V8 51

Gardner, Goldie 11, 85, 88
Garner, H.M. 181
Gawn, Richard 80, 81, 82, 83, 98, 99, 117, 126, 129, 130–132, 133, 135, 147, 149, 172
Gemmell, Jock 252, 258–259, 304
Gilmore Oil 68, 71
GKN (Guest, Keen & Nettlefolds) 188
Glasgow Herald 271, 273
Glendenning, Raymond 276
Glen Urquhart Community Association 325
Goodwood 252
Greavette Boats, Ltd 51
   Sheerliner 51
   Streamliner 51
Greecen (logging system) 278–279
Greening, Harry 80, 82

Hagg, Arthur 180
Halford, Frank 56, 84, 94, 95, 96, 97, 98, 102, 107, 110, 145, 147, 182, 207, 225
Halliwell, Victor 46
Hall-Scott Motor Company 55
Hanning-Lee, Frank 288, 289, 322, 323

Hanning-Lee, Stella 288, 289, 322, 323
Harmsworth Trophy 63, 288
Harris, Fred 11
Haslar — see 'Admiralty Experiment Works'
Henderson, Deryck Farquharson 41
Higham Special (car) 24
Hives, Ernest (Lord) 31, 290
Hobbs, Mr (of Thomson & Taylor) 66
Hodge, M.D. 281, 285
Holloway, Dudley 42
Holloway, Eileen — see 'Cobb, Eileen Lucy (sister)'
Hooker, Stanley 321
Hooper, Maurice 125, 145, 157, 167, 168, 170, 171, 172
Hudson (cars) 30, 87, 88, 108, 110, 128, 156, 159
   Terraplane 87, 88

ICI (Imperial Chemical Industries) 123, 190
Invicta (cars) 30
Isabelle (boat) 242, 250, 312

Jaguar (cars) 42
Jetex (rocket motors) 115
Jetstar (boat) 53
Johnson, Leslie 318, 321
Jones, Hughie 96, 201, 204, 208, 212, 214, 217, 220, 221, 223, 224, 227, 242, 248, 249, 285, 337
Jones, Tudor Owen 'Ted' 51–52, 60, 61
Joslin, Sally — see 'Railton Joslin, Sally'

Karunna (boat) 240, 241, 242
Kine Engineering 289
Kleboe, Raymond 243, 273–274
Kongsberg Maritime 330, 332, 333, 334
   Hugin 334
   Munin 332, 334, 336

Lake Eyre (Australia) 291
Lake Garda (Italy) 62
Lake Hallwyl (Switzerland) 47, 49
Lake Maggiore (Italy) 48
Lake Windermere (Lake District) 47, 166, 176

Lazenby, Ernie 294
Leach, Harry 61
Leyland Motors 21, 22, 23
   Eight 21–23, 24, 30
   Leyland-Thomas 24, 35
Liberty (engines) 24
Liberty Ship 55
Lillicrap, Sir Charles S. 90, 146
Lincoln (cars) 94
Loch Lomond (Scotland) 166, 170
Loch Ness (Scotland) 7, 9, 155, 176, 197, 200–239, 240–252, 253–287, 288–315, 317, 325, 326–327, 329, 330–338, 340, 341, 342
   Monster 42, 334
   Temple Pier 200, 202, 203, 204, 242–244, 245, 249, 251, 262, 277, 280, 286, 340, 342
Loch Ness Project 330, 332
Loebl, Henry 221
London Motor Show 1920 21
Lorant, Stefan 273
Lotus (Formula 1) 76, 222
Low, David 259, 262, 306
Lucas (electrics) 31, 272
Lucas, Oliver 31
Ludvigsen, Karl 9, 11
Lycoming (engines) 51
Lydall, Frank 233, 236–237, 250, 262, 263, 273, 285

Macdonald, Dr William 250
Macklin, Noel 30
MacRae & Dick 272
'Mae West' (lifejacket) 197, 248
Maina, Joseph 27
Maloney, Bill 286–287, 291, 294, 296, 297
Marine Aircraft Experimental Establishment 180, 181, 279, 304
Marine Motoring Association 192, 193, 194, 280
Masters, Bill 66
Maureen Mhor (boat) 235, 240, 241, 242, 244, 245, 246, 248, 249, 250, 252, 259, 262, 263, 266, 269, 277, 280, 281, 282, 284, 285, 286, 305, 307, 308
Mayne, Philip 243, 285, 307
Mays, Raymond 30, 87

349

McNaughton, Mr (of MacRae & Dick) 272
Meacham, Larned 76–78
Meacham, William 76–78
Meldrum brothers 289
Menzies, Alec 200, 202, 226, 234, 240, 242, 251, 269, 270–271, 273, 277, 279, 333, 339–340
Menzies, Dorothy 339
Menzies, Gordon 273, 277–278, 333
Menzies, Jimmy 339
Menzies, Margaret 'Peggy (née Montgomery) 339
Mephistopheles (car) 35, 36
Mercedes (cars) 24
Messerschmitt
   Me 263 '*Komet*' 104
MG
   EX135 (record car) 85, 88
Milledge, Zillwood 'Sinbad' 86–87
Minerva (cars) 33
Minett-Shields (boatbuilder) 51
Ministry of Supply 104, 145, 182
Miss Canada (hydroplane) 75, 80, 82, 84, 85, 108
   Miss Canada IV 83, 106
Miss England (record boat) 45
   Miss England II 45–46, 48
   Miss England III 97, 166, 170
Miss America X (record boat) 45
Mobil (fuel and oil) 70, 71
Monza 24
   1924 Italian Grand Prix 24
More, Sir Thomas 63
Morris Minor 165
Moss (gearboxes) 23
Motor Boat and Yachting 74–75, 155, 158, 177, 180, 224, 310–311
Mouat, Nick 328–329
Movietone News 197, 260–261, 277, 303, 306

Napier (aero engines) 37, 71
   Lion 63, 65
Napier-Railton (car) 37–38, 39, 63, 76, 234, 320
National Geographic 9, 334
National Maritime Museum 52
National Physical Laboratory 64, 328

Newton, Julie (*née* Bennetts) 329
Newton, Len 329
Noble, Richard 9, 328–329, 342–343
Norris Bros (engineering consultancy) 289, 291, 292–293, 296, 298, 313, 322
Norris, Eric 289
Norris, Ken 181, 287, 288–289, 291, 296, 297, 299, 322, 323, 324, 325, 343
Norris, Lewis 'Lew' 181, 288, 289, 297, 322, 325
Notley, Harold 51
Nye, Doug 42, 43, 276

Oltranza Cup 62
Owen, Alfred 31
Owen Organisation 31

Packard (aero engines) 45
Panhard (cars) 33, 36
Parry Thomas, John Godfrey 21–28, 30, 31, 35, 36, 44, 222
Pathé News 197, 237, 252, 258, 304
Patiala, Maharajah of 22
Patience, Hugh 206, 226, 240, 241, 242, 270, 279, 305
Pendine Sands, Wales 24–28
Perreit, John 324
Peters, Captain Harold 40
Phillips-Birt, Douglas 287
Picture Post 273
Posthumus, Cyril 42
PYE Electronics (radio equipment) 205, 216, 241, 244, 263, 271, 281
   Dolphin intercom 281

Queen's Commendation for Brave Conduct 326
Queen Elizabeth The Queen Mother 236, 237
Queen Elizabeth II 241

Railton (cars) 30, 87, 88
Railton, Audrey (wife) 164, 321
Railton, Charles Withington (father) 21
Railton, Charlotte Elizabeth (mother) (*née* Sharman) 21
Railton Joslin, Sally (daughter) 7–9, 11, 12, 164

Railton, Reid 7–9, 11, 12, 17, 18, 19–24, 25, 27–29, 30–31, 34, 37, 39, 42, 44, 45, 46, 47, 48, 49, 50, 51, 52, 54, 55, 56, 57, 58, 60, 61, 63–66, 68–72, 73–88, 89–136, 137–164, 165–177, 180–181, 183–184, 185, 189, 192, 193, 194, 196, 197, 200, 201, 204, 205, 206, 207, 210, 211, 214, 216, 218, 220, 221, 222, 224–225, 226, 227, 229, 234, 236–237, 238, 239, 252, 267, 273, 274–275, 276, 279, 286, 288, 289, 290, 291, 297, 299, 300, 305, 308, 313, 315, 316–325, 328, 329, 331, 335, 337, 342, 343
Railton, Tim (son) 7, 164
Railton Special (record car) 44, 64–72, 76
Railton Mobil Special (record car) — see '*Railton Special*'
Randall, Sid 61
Rees, Bill 205, 208, 216, 241, 244, 245, 263
Reliant Robin 303
Richmond, Duke of 252
Riddell-Black, Robert 259
Riley (cars)
   Nine 28
   Brooklands Model 38
   One-Point-Five 8
Riley, Victor 38
Robison, Howard 318
Rolls-Royce (cars) 21
Rolls-Royce (engines) 31, 54, 56, 57, 107, 114, 290, 317, 318, 321, 342
   R-type 31, 45, 48, 54, 56, 61, 63, 70, 267
   Number R39 61
   Griffon 56, 289
   Merlin 54, 56
   Avon jet engine
      RA3 Mk 101 322
Royal Aircraft Establishment 90
   Rocket Propulsion Division 90, 104, 114
Royal Air Force 38, 269, 304, 318
Royal Automobile Club 72, 193
Royal Flying Corps 22
Royal Navy 342
Rubery Owen 24
Rudd, Tony 31

INDEX

SAAB Tunnan 182
Saunders-Roe 46, 54, 84, 95, 97, 159, 183, 186, 317, 318, 322
Sayres, Stanley 17, 50, 51, 52, 60, 61
Schneider Trophy 145, 180, 304, 305
Scott Russell, John 279
Segrave, Henry (Sir) 17, 25, 27, 34, 45–46, 73
Segrave Trophy 72
Shell (fuel and oil) 317, 318, 321
Shell-Mex (fuel and oil) 25
Shine, Adrian 280, 330, 332, 333
Shipbuilding & Shipping Record 287
Ships
  RMS *Caronia* 160, 239
  RMS *Queen Mary* 145
  SS *Lafonia* 276
  SS *Wedia* 148
Sidgreaves, A.F. 54
Silverstone 236
Simpson, Sammy 94
Slo-mo-shun II (hydroplane) 52
Slo-mo-shun IV (hydroplane) 50, 52, 60, 61, 152
Smith, A.G. 180–181
Smith, R.E. 194
Soliton effect 279, 280, 312
Southport Sands 25
Spirit of Australia (record boat) 6, 181, 343
Standard Motor Company 31
Spurrier, Henry 23, 120
Sultan, Laurie 72, 273
Sunbeam 350HP (record car) 25
Sunbeam 1000HP (record car) 27
Supermarine 151, 183
  S6 305
  S6B 305
  Swift 151
Sweeney, Diana 'Bunny' 7–8, 42

Taylor, Ken 28–29, 44, 68, 163–164, 181, 218, 221, 239, 324
Templar Films 259

Test models 90–92, 98, 108, 115–116, 119–121, 123–124, 125–133, 135, 137, 138, 140–144, 146, 147, 151, 153–154, 155, 157, 162–163, 166–167, 169, 180, 215
CJK 91, 137, 139
CDQ 139
CJQ 151, 153, 157
The Autocar 41
The New York Times 249
Thomas Inventions Development Co. Ltd. 24, 28, 31
Thomas, John Godfrey Parry — see 'Parry Thomas, John Godfrey'
Thomas, René 36
Thompson, John 150
Thompson, Mickey 71
Thompson Motor Pressings Ltd 66, 150
Thomson, Ken 21, 23, 24, 25, 27, 28, 44, 52
Thomson & Taylor 21, 28–29, 30, 36, 37, 52, 56, 64, 70, 73, 74, 79, 85–86, 88, 90, 92, 93, 114, 125, 137, 147, 218, 219, 222, 235, 295, 313, 324
Thornycroft 94, 96, 97, 100, 102, 103
Thrust SSC (record car) 329
Tirpitz (warship) 124
TOPS III (hydroplane) 51
Thunderbolt (record car) 63, 65, 68, 202, 234
Thwaites, Goffy 61
Tyndall, D. 268–269
Tyson, Geoffrey 44

Ullswater (Lake District) 325
Ulster TT, 1929 38
Union International Motonautique (UIM) 17, 280

Valveless (cars) 33
Van Patten, Douglas 51, 52, 74, 75, 82, 83, 84, 89, 90, 92, 93, 106, 108, 109, 110, 111, 113–114, 115, 116, 117, 118, 121, 123, 125, 126, 127, 129, 130–131, 133, 135, 137, 151, 188, 328, 331
Vauxhall (cars)
  TT 38

Ventnor Boat Works 49
  Ventnor hull 49–50, 51, 58, 94
Vickers Ltd 29, 47, 53, 64
  Wellington bomber 42
  Wind tunnel 64
Villa, Leo 29, 49, 59, 60, 61, 98, 202, 322, 324
Vosper 17, 47, 51, 52, 56, 58, 60, 73, 74, 76, 79, 80–81, 82, 84, 86, 88, 89–93, 94, 95, 96, 97, 98, 99, 100, 101, 103, 108, 109–112, 114, 117, 118, 120, 122, 126, 133, 134, 135, 136, 137, 142, 144, 145, 146, 147, 148, 151, 152, 153, 156, 158, 160, 161, 163, 164, 166, 168, 170, 171, 172, 173, 175, 176, 177, 182–191, 194–198, 201, 207, 210, 211, 214, 215, 221, 222, 224, 227, 230, 233, 234, 235, 236, 237, 239, 242, 245, 270, 271, 272, 274, 277, 284, 285, 286, 288, 295, 296, 300, 302, 307, 313, 318, 320, 339
Vosper Thornycroft 12

Wakefields — see 'Castrol'
Wallace, Craig 330–338
Walter rocket motor
  Type 509A 104
Warby, Ken 6, 181, 291, 343
Warde, Richard 35
Westcott (rocket motors) 172
Wheeler, Ken 291, 296, 297, 299
White Hawk (record boat) 288, 323
Whittle, Frank (Sir) 31
Whyte, Andrew 42
Whyte, Constance 42
Wilcocks, Michael 46
Wilkins, Richard 'Dick' 316, 318–321
William Mallinson & Sons 183
William O. Smith & Associates 51
Williams (Formula 1) 76
Winn, Wing Commander C.V. 180–181, 304
Wokington Plastics 190
Wood, Gar 45

Yachting World 178–179

Zborowski, Count Louis 24

351

AUTHOR

Steve Holter's life-long association with all things automotive, both as a hobby and a profession, and as a driver and an engineer, made it inevitable that his interests would gravitate towards record-breaking. Mixing his professional experience — including television research and crash investigation — with knowledge gained while working with his friend and mentor, *Bluebird* designer Ken Norris, as well as numerous interviews with many others involved in record-breaking, he is in a unique position to unravel the true story of *Crusader*. He has written one previous book, *Leap into Legend: Donald Campbell and the Complete Story of the World Speed Records* (Sigma Press, 2002). He lives in Belves, France.